Author's Acknowledgements

I would like to express my sincere thanks to the friends at my workshop for editing, trailing, advising and reporting on the materials.

S K So
Chief Author

The Publisher welcomes any comments from readers. Please email us at sosir2006@hotmail.com. For school order, please contact the Marketing Department on (852) 9034 9915.

LCCI初級及中級
會計精讀 ✓ PASSED
必勝攻略

作者	蘇斯杰 (So Sze Kit), PhD (Cand.), MBA (OUHK), PGDE (CUHK)
出版	超記出版社 (超媒體出版有限公司)
地址	荃灣海盛路 11 號 One MidTown 2913 室
電話	(852) 3596 4296
電郵	info@easy-publish.org
網址	http : //www.easy-publish.org
香港總經銷	香港聯合物流有限公司
上架建議	商業會計
ISBN	978-988-8700-15-8
定價	HK$98

U0152213

Printed and Published in Hong Kong
版權所有 · 侵害必究

如發現本書有釘裝錯漏問題，請攜同書刊親臨本公司服務部更換。

會計這一科真的值得讀嗎？

為何要修讀 LCCI 會計課程

    LCCI 為 London Chamber of Commerce & Industry 的簡稱，已經成立了超過一百多年。LCCI 考試提供 60 種以上的專業考試，而其中會計是最為香港人所熟悉的。LCCI 前主席 Sir James Duncan 曾表示，得到 LCCI 會計資格，猶如得到 Passport to Employment（就業通行証）。如留意香港報章的求職版，最多的職位空缺是業務代表，其次就是會計、財務、金融類等行業，所以擁有 LCCI 會計證書是非常重要的。

    另一方面，從投資角度看，考取 LCCI 會計證書亦是一個合適的投資。以香港職場的文員為例，持有 LCCI 會計證書除了入職會較優先外，起薪點亦平均較高，約 300-500 元。

考取 LCCI 所需的開支：

修讀會計課程	$1000
會計教科書（多本）	$200
考試費	$400
總投資數	$1,600

    如果入職多了 $300，那麼，5 個月便回本了，第六個月便是賺。比投資開店做生意還要快，作為精明的你，又怎會放過這機會？

報考方法

季度	報考日期	考試日期
Series 2	Oct — Nov	下年 Apr — May
Series 3	Dec — Jan	下年 Jun — July
Series 4	Jun — Jul	同年 Nov — Dec

· 報表格可向就讀學校提出

· 可向 LCCI 香港辦事報告，查詢電話：(852) 3102 0100

· 如有疑問，也可向作者蘇老師查詢：sosir2006@hotmail.com

報考須知

· 每季考試均有兩星個日期，考生無權選擇應考日期，而日後不能改的，考生必須準備讓兩天時間。

· 准考證會於開考約前四星期送到報名學校，考生請記緊自行致電報名學校查詢。

· 收到准考證後留意姓名、身份證號碼是否正確。

LCCI 考試之準備

◎ 溫習重點

LCCI Level 1 重點：
· Bank Reconciliation Statement
· Petty Cash Book and the Imprest System.
· Capital and revenue expenditure.
· Adjusting for accruals and prepayments.
· Depreciation of fixed assets.
· Bad debts and provision for doubtful debts.
· T, P&L a/c 及 Balance Sheet

LCCI Level 2 重點：
· Partnership
· Limited Company
· Single Entry and Incomplete Record
· Trading, Profit & Loss Account
· Balance Sheet
· Income & Expenditure a/c

◎ 時間分配

· 考生可用大約 35 分鐘去完成一題。
· 不要放棄作答任何一條試題，要盡量分配時間作答所有題目。

◎ 100 天前準事宜

· 計數機印
· 考生應至少溫習三年以上的歷屆試題，才能真正掌握出題趨勢及作答要求。
· 溫習 Past Paper 時，不能只看答案，記著要白紙黑字做出來呢！
· 溫習如有不明白之處，立即向作者蘇老師提問：sosir2006@hotmail.com
· 向公司申請考試當日請假半天。
· 事前要探路，了解如何到達試場，除了的士外，還有何途徑。

◎ 試前一晚的準備

1. 文件
   · 准考證 (仔細檢查檢查日期)
   · 身份證
2. 文具安排
   · 鉛字筆 (2 或更多)
   · 透明間呎 (2 或更多)
   · 計數機 (已蓋印)
   · 改錯帶
   · 螢光 / 顏色筆
   · 鉛筆
3. 運輸安排
   · 前往路線 (多於一個的選擇)

- ・硬幣
- ・八達通
4. 食品安排
   - ・晚餐
   - ・蒸餾水
5. 衣服
   - ・外套
   - ・運動鞋 (方便「追」巴士)
   - ・手巾 / 紙巾
   - ・髮夾
6. 醫學安排
   - ・Panadol
   - ・香薰油或白花油 (提神用的)

◎ 注意事項

- ・先花三數分鐘時間來審核試題
- ・選擇最有把握的題目作答，然後才作答較沒把握的題目。
- ・仔細閱讀 REQUIRED 的部分
- ・Underline 關鍵及重要字 (利用不同顏色)
- ・所有作答均要清楚及整齊
- ・緊守 35 分鐘一條的時間控制策略，不能超時。
- ・嘗試回答所有問題
- ・一條都不能少

◎ 分數表現

- ・Pass → 50%
- ・Credit → 60%
- ・Distinction → 75%

◎ 評分準則

- ・答對地方便有分
- ・如果計算錯誤，但計算步驟清楚列明，有機會得到部分分數

◎ 後記

- ・充份準備才能使你的壓力下降！

如想有 Credit 或以上，只靠溫習課本是不足夠的，必須要大量操練試題。你試過操練時遇到問題不知如何是好嗎？本書作者從事會計培訓多年，深明問題之所在，為了協助勤力的學生，現特設網絡支援服務，包括：電郵問書服務、最近模擬試題、模擬試題批改等等。如果你對考試範圍有疑問，可將題目電郵至 sosir2006@hotmail.com 向作者發問，之後便會有專人回覆及跟進，為你解決疑難。

LCCI 升學途徑

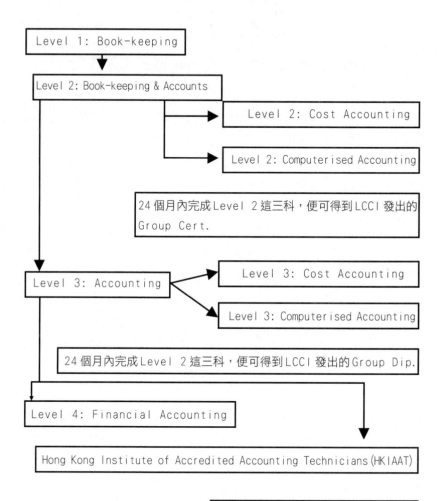

Level 1: Book-keeping

Level 2: Book-keeping & Accounts

Level 2: Cost Accounting

Level 2: Computerised Accounting

24 個月內完成 Level 2 這三科，便可得到 LCCI 發出的 Group Cert.

Level 3: Accounting

Level 3: Cost Accounting

Level 3: Computerised Accounting

24 個月內完成 Level 2 這三科，便可得到 LCCI 發出的 Group Dip.

Level 4: Financial Accounting

Hong Kong Institute of Accredited Accounting Technicians (HKIAAT)

Paper 1: Financial Accounting

如 LCCI Level 2 考試中，得到 Credit 或以上之成績，可申請豁免這一張卷。

Level 1

《LCCI 初級會計 (Level 1) 精讀備試天書》

Content 目錄 ////

本書附送 LCCI 增值錦囊，請到以下網址下載：
http://www.systech.hk/lcci-note.rar

Chapter 1

Introduction and Day Books

Content

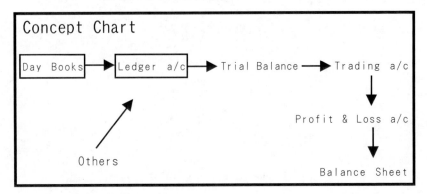

Concept Chart

Day Books → Ledger a/c → Trial Balance → Trading a/c

Others

Profit & Loss a/c

Balance Sheet

[01] What is Accounting?

Accounting Concerned

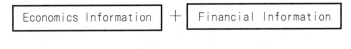

| Economics Information | + | Financial Information |

Function

| Better Decision | + | Budget for future |

[02] Process of Accounting

In general, the process of accounting can be divided into 5 stages:

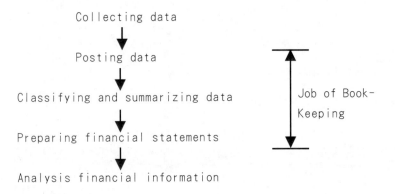

Collecting data
↓
Posting data
↓
Classifying and summarizing data
↓
Preparing financial statements
↓
Analysis financial information

Job of Book-Keeping

A professional accountant has the responsibility of carrying out all of the above stages.

[03] What is book-keeping?

◎ to measure accounting data of an undertaking
◎ to communicate the business results to the interested parties such as board of directors.

[04] Users of Accounting Information

a.Owner(s) of the business

◎ They want to know the business is profitable or not.
◎ They want to know what the financial resources of the business are.

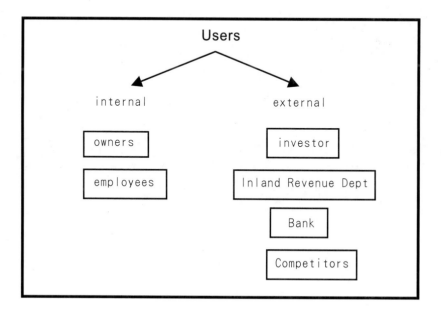

b.A prospective investor 準投資者

When the owner wants to sell it, the potential investor must want to see such information.

c.The bank

If the owner wants to borrow money for the business, then the bank will need such information.

d.The Inland Revenue Department 稅務局

They need it to calculate the taxes payable.

e.Competitors

Competitors are interested to evaluate their own efficiency and to monitor competition.

f.Employees

They concern on their job security and remuneration.

[05]　Accounting Equation

The whole idea of financial accounting can be summed up by the accounting equation which is shown as follows:

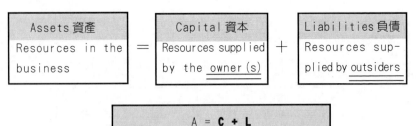

After making profit:

Assets = Liabilities + Capital + Income − Expenses

Assets + Expenses = Liabilities + Capital + Income

10

a. Assets (A)

They are economic resources, which help to make profit

Fixed Assets FA

◎ are assets to be kept as such for more than one year

e. g. Land and Buildings, Furniture & Fitting, Office Equipment

Current Assets CA

◎ are assets which change from day to day

e. g. Stock, Debtors, Bank, Cash

b. Liabilities (L)

They are debts or obligations owed by the business to outside parties.

Long-term liabilities / Amount due more than one year LL

◎ are those liabilities which have to be paid more than one year

e. g. Bank loans, 5% Debentures

Current liabilities / Amount due within one year CL

◎ are those liabilities which have to be paid within one year.

e. g. Creditors, Bank overdrafts, Accrual Expenses

c. Capital (C)

The total of resources supplied by owner. It is a liability owed to its owners.

d. Income (I)

It can increase the capital.

e. g. Sales, Rent received, Interest received

e. Expenses (E)

Expenses refer to all expenditures incurred in the usual courses in earning income.

(i) Trading Expenses e. g. Purchases, Carriage inwards

11

(ii) Administrative Expenses e.g. Rent and rates

(iii) Selling and Distribution Expenses e.g. Carriage outwards, Commission

(iv) Financial Expenses e.g. Bank interest

Classwork 1

For each of the following items, put the symbol in the space provided to show the major accounting category.

Symbols: Assets: A; Capital: C; Liability: L

1. Capital introduced_____C_____
2. Premises _____
3. Stock _____
4. Loan from Chan _____
5. Trade debtors _____
6. Loan to May _____
7. Creditors _____
8. Motor vehicles _____
9. Deposit at HSBC Bank _____
10. Invest in shares of Tom Online _____

Classwork 2

Supply the missing amount of the following cases:

	Assets $	Liabilities $	Capital $
a)	50,000	19,000	_____
b)	_____	19,300	54,400
c)	38,000	_____	28,000
d)	219,500	25,500	_____
e)	78,000	_____	40,000

12

[06] The Sales Journal and The Sales Ledger

a.Sales

Cash sales	Credit sales
When cash receives for selling goods immediately, there is no need to enter such sales in the sales journal.	In many businesses most of the sales will be made on credit

b.Sales Journal (Sales Day Book)

From the copies of the sales invoices, the book-keepers record data to their sales journals.

c.Sales Ledger

It is used for recording customers' (debtors') record.

d.The Sales Journal and the Sales Ledger

Posting credit sales ONLY

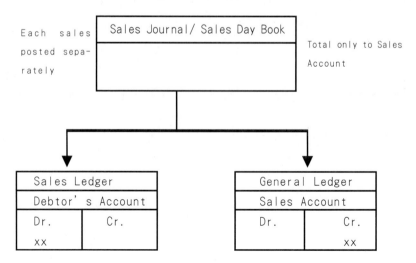

13

Classwork 3

Mr. Mak has the following sales for the month of March 20X7.

Mar 1   Sold goods on credit to Lily Ltd. for $40.

Mar 5   Sold goods on credit to Lily Ltd. for $200 less 20% trade discount.

Mar 25   Credit sale to Betty Ltd. for $100.

Mar 27   Cash sales $100 to Betty Ltd

Mar 30   Sold goods on credit to ABC Company for $40 less 10% trade discount.

Sales Journal / Sales Day Book

Date	Particulars	F	Amount ($)	Total ($)

[07] The Purchases Journal and the Purchases Ledger

a.Purchases

(i)Cash purchases

When cash pays for goods immediately, there is <u>no</u> need to enter such purchases in the purchases journal.

(ii)Credit purchases

In many businesses most of the purchases will be made on credit for immediate cash.

b.The Purchases Journal

It is used for recording all credit purchases.

c.The Purchases Ledger

It is used for recording suppliers' (creditors) record.

d.The Purchases Journal and the Purchases Ledger

Posting Credit Purchase ONLY

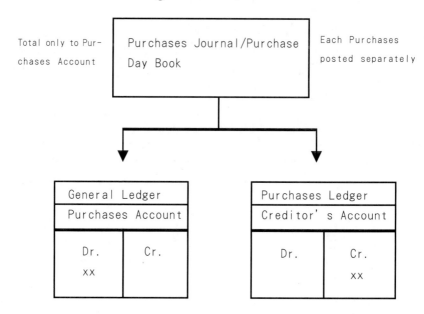

Classwork 4

Mr. Tang has the following transactions during the month of March 20X7 :

Mar 4	Purchased goods on credit from Yuen Ltd. for $1,400 less 5% trade discount.
Mar 15	Purchased goods on credit from Yuen Ltd. for $1,800 for 10% trade discount.
Mar 26	Purchased goods on credit from Main Main Ltd. for $3,100.
Mar 29	Purchase goods from Main Main Ltd $2,600 by cheque

Purchase Journal / Purchase Day Book

Date	Particulars	F	Amount ($)	Total ($)

[08]   The Returns Inwards Journal

a. Returns Inwards and Credit Notes

(i) Returns inwards journal

It is used to record goods returned to us

(ii) Credit notes

It is a document sent to a customer by the business, showing allowance given by supplier in respect of unsatisfactory goods.

Posting Return Inwards

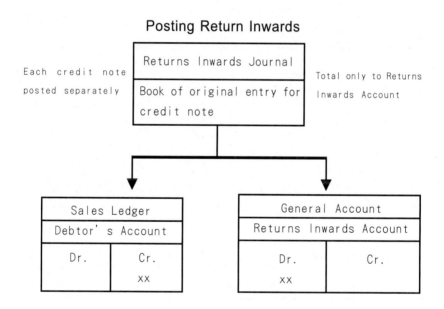

Each credit note posted separately

Returns Inwards Journal

Book of original entry for credit note

Total only to Returns Inwards Account

Sales Ledger	
Debtor's Account	
Dr.	Cr.
	xx

General Account	
Returns Inwards Account	
Dr.	Cr.
xx	

Classwork 5

An example of a returns journal:

Returns Inwards Journal

Date	Particulars	Folio	Amount
20X7			$
Sep 2	Chun Store	SL 1	400
17	Peter Ltd.	SL 2	120
19	Amy	SL 3	290
29	Super Store	SL 112	160
30	Total transferred to Returns Inwards Account	GL 1	972

Return Inwards

Debtor

[09]   The Returns Outwards Journal

a. Returns Outwards and Debit Notes

(i) Returns outwards journal

It is used to record goods returned to our suppliers.

(ii) Debit Notes

It is a document sent by a business to their supplier, giving details of a claim for an allowance in respect of unsatisfactory goods.

Posting returns outwards

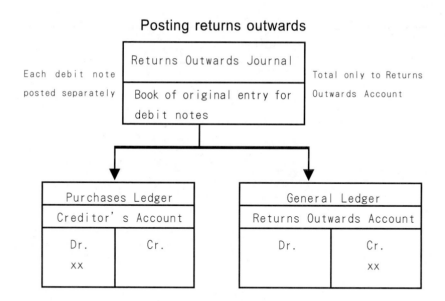

Classwork 6

An example of a returns outwards journals:

Returns Inwards Journal

Date	Particulars	Folio	Amount
20X7			$
Sep 11	John	PL 29	180
16	X Ltd.	PL 46	100
28	Silly	PL 55	30
30	Jimmy	PL 4	360
30			
	Total transferred to Returns Outwards Account	GL 2	690

Return Outwards

Creditors

[10] Examination

2003		S3	
2004		S3	
2005	S2	S3	S4
2006	S2		

Chapter 2

Double Entry System

Content

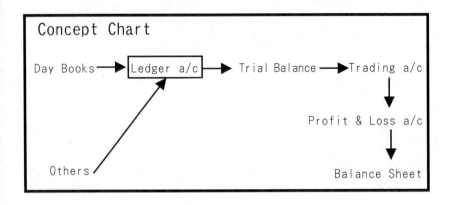

Concept Chart

Day Books → Ledger a/c → Trial Balance → Trading a/c
                                              ↓
                                          Profit & Loss a/c
                                              ↓
Others   Balance Sheet

[01] Nature of a transaction

Events that result in such changes are known as transactions.

[02] Double entry system

Account is part of double entry records, containing details of transactions for a specific item.

Double entry bookkeeping is a system by which each transaction is entered twice.

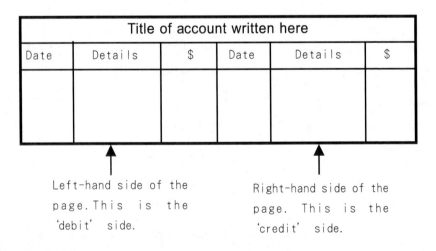

Title of account written here					
Date	Details	$	Date	Details	$

Left-hand side of the page. This is the 'debit' side.

Right-hand side of the page. This is the 'credit' side.

[03]   Double entry rules

	DR	CR
Income (I)	−	+
Expenses (E)	+	−
Assets (A)	+	−
Liabilities (L)	−	+
Capital (C)	−	+

[04]   Stock movements

Account	Reason
Purchases account	For the purchase of goods for resale
Sales account	For the sale of goods in the normal of business
Returns Inwards account	For goods returned to the firm by its customers
Returns Outwards account	For goods returned by the firm to its suppliers

[05]   Purchases of stock on credit

Example 1 : 1 August 20X7.   Goods costing $1,000 were bought on credit from Super Store.

23

```
_____|_____
                               |
                               |
                               |
                               |
                               |
_____|_____
                               |
                               |
                               |
                               |
                               |
```

[06]   Purchases of stock for cash

Example 2 : 2 August 2003.   Goods costing $200 was bought,
cash being paid for them immediately.

```
_____|_____
                               |
                               |
                               |
                               |
_____|_____
                               |
                               |
                               |
                               |
                               |
```

[07]   Sales of stock on credit

Example 3 : 3 August 2003.   Sold goods on credit for $250 to Lee.

[08]   Sales of stock on cash

Example 4:   4 August 2003.   Goods were sold for $500, the cash being received at once upon sale.

[09] Returns Inwards

Example 5:   5 August 20X7.   He returned goods, which had previously been sold to Lo for $300.

[10] Return Outwards

Example 6:   6 August 2003.   Goods previously bought for $900 were returned by the firm to K. So.

[11]　Revenues

Example 7:　5 June 20X7. We let someone else use our premise and received rent of $40 by cheque.

[12] Drawings

Drawings consist of cash or goods taken out of a business by the owner for his private use.

Example 8:   25 August 2003.   Proprietor took $100 cash out of the business for his own use.

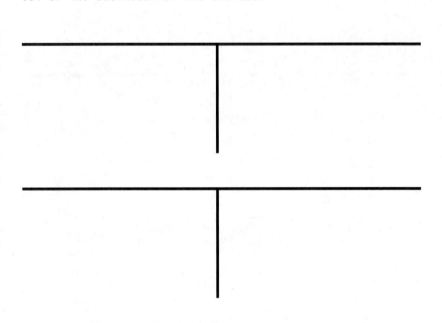

[13] Payment of expenses

Example 9:   26 Aug 2003, we paid $2,500 of rent by cheque.

[14] Assets Increase

Example 10: Buy car $10,000 from ABC Ltd $10,000, we paid $1,500 for deposit, the balance will pay one month later.

[15]　Capital Movement

Capital

Drawings	Bal b/d
Net Loss	Cash / Bank
Bal c/d	Net Profit

[16]　Balancing

It means locating the difference between two sides on the smaller side, so as to equalize the totals on each side of the account.

Example 11

Debtors				
2005	$	2005		$
Jan 1　Sales	1,000	Jan 2　Bank		250
Jan 3　Sales	450	Jan 4　Cash		20
		Jan 31　Bal c/d		

[17]　Examination

2003		S3	S4
2004	S2		S4
2005	S2	S3	
2006	S2		

Chapter 3

Trial Balance

Content

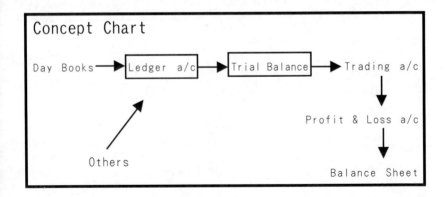

Concept Chart

Day Books → Ledger a/c → Trial Balance → Trading a/c

→ Profit & Loss a/c

→ Balance Sheet

Others

[01] Balancing off Accounts

At the end of the accounting period, all accounts should be balanced off.

The balance over the totals is called as balance carried forward (bal c/f) or balance carried down (bal c/d). The balance below the total is called as balance brought forward (bal b/f) or balance brought down (bal b/d).

Example 1:

Debtor - Mei

2004		$	2004		$
Mar 3	Sales	158	Mar 18	Bank	158
Mar 15	Sales	206			
Mar 28	Sales	118	Mar 31	Bal c/d	324
		482			482
Apr 1	Bal b/d	324			

32

Example 2:

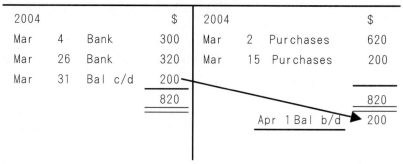

Creditor - Lee

2004			$	2004			$
Mar	4	Bank	300	Mar	2	Purchases	620
Mar	26	Bank	320	Mar	15	Purchases	200
Mar	31	Bal c/d	200				
			820				820
				Apr	1	Bal b/d	200

[02] Different cases of balance c/d

1. Cases of balancing the account $\boxed{Dr = Cr}$

Peter

2006			$	2006			$
Jan	1	Sales	200	Jan	2	Cash	100
	4	Sales	500		4	Cash	600
					31	Bal c/d	
			700				700

2. Cases of balancing the account $\boxed{Dr > Cr}$

Peter

2006			$	2006			$
Jan	1	Sales	200	Jan	2	Cash	100
	4	Sales	500		4	Cash	450
					31	Bal c/d	
			700				700

3. Cases of balancing the account $\boxed{\text{Dr} < \text{Cr}}$

Peter

2006			$	2006			$
Jan	1	Sales	100	Jan	2	Cash	100
	4	Sales	500		4	Cash	600
	31	Bal c/d					
					31	Bal c/d	
			700				700

Classwork 1

Jun	1	Cash sales $1,000
	9	Credit Sales $1250
	12	Cash Purchase $200
	23	Cash Sales $1,000
	30	Paid Rent $50

Required:

Prepare the Cash a/c and balance off

Cash

2006		$	2006		$
Jun			Jun		

[03] Format of Trial Balance

At the end of the accounting period, all accounts should
be balanced off or closed. All the balances are extracted to
a trial balance.

34

K. Poon

Trial Balance as at 31 May 2001

	Dr.	Cr.
	$	$
Bank	xxx	
Cash	xxx	
Yip		xxx
Mok	xxx	
Capital		xxx
Purchases	xxx	
Sales		xxx
Returns inwards	xxx	
Returns outwards		xxx
Machinery	xxx	
	xxxx	xxxx

Dr side	Cr side
A, E	C, L, I

Classwork 2

Purchase

	$		$
Cash	100		
Cash	20		

Debtor

	$		$
	700		

Sales

	$		$
		Cash	250
		Cash	500
		Debtor	700

Cash

	$		$
Bal b/d	100		

Balance the above ledger account and then prepare trial balance.

Trial Balance

	Dr.	Cr.
	$	$
Purchase		
Sale		
Cash		
Debtor		

[04]   The use of Trial Balance

There are several uses to which a trial balance may be put. These are:

◎ To check that the books 'balance'.

◎ To ascertain the net amount of the error(s), if any have been made, as represented by the difference between the two sides.

[05]   Types of accounts

(i)   Personal account − Debtors, Creditors, Capital
(ii)   Nominal account − Incomes, Expenses
(iii)   Real account − others

[06]   Examination

2003		S3
2004		
2005	S2	

Chapter 4

Final Accounts

Content

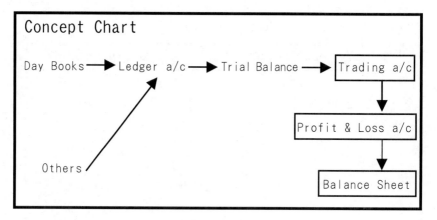

[01] Gross Profit and Net Profit

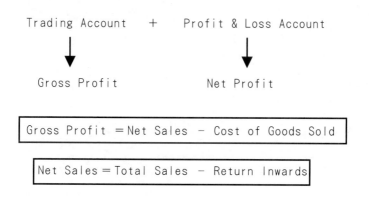

Gross Profit = Net Sales − Cost of Goods Sold

Net Sales = Total Sales − Return Inwards

[02] Equation of "Cost of Goods Sold"

Cost of Goods Sold
= Opening Stock + (Purchase − Return Outward + Carriage Inwards)
  − Closing Stock

[03] Items of Trading Account

Items of " T" → Related to 'Stock' !

[04] Vertical style for trading accounts

Trading Account for the year ended 31 December 20X7		
	$	$
Sales		3,850
Less : Cost of goods sold:		
Opening Stock	100	
Purchases	2,900	
Closing stock	(400)	2,600
Gross profit		1,250

Classwork 1

Sale	200
Opening Stock	1500
Purchase	700
Closing Stock	1000

[05] Vertical style for profit and loss accounts

NP = GP + other income - expenses

Profit and Loss A/C for the year ended 31 December 20X7		
	$	$
Gross profit		1,250
Less: Expenses:		
Rent	240	
Lighting	150	
Heating		
Net profit		800

Classwork 2

Rent   200
Rate   100

[06]   Vertical style for trading and profit and loss A/C

Trading and Profit and Loss A/C for the year ended 31 Dec 20X7		
	$	$
Sales		3,850
Less: Cost of goods sold:		
Purchases	2,900	
Closing stock	(300)	2,600
Gross profit		1,250
Less: Expenses:		
Rent	240	
Lighting	150	
Heating	60	450
Net profit		800

[07]   Classification

Item	Class	Definition	Examples
Assets	Fixed Assets	· Long life use · Not for resale	Land, buildings
	Current Assets	· Short life use · For resale	Stock, debtors
Liabilities	Long Term Liabilities	· More than one yr	Bank loan
	Current Liabilities	· Less than one yr	Creditors

[08] Vertical Style balance sheets

$$A = C + L$$

A   C + L

Balance Sheet as at 31 December 20X7

	$	$	$
Fixed Assets	Cost	Acc. Depr	NBV
Fixtures and fittings	500	0	500
	500	0	500
Current Assets			
Stock	300		
Debtors	680		
Bank	1,510		
Cash	20	2,510	
Less Amount due within one year			
Creditors		910	
Net Current Assets (CA − CL)			1,600
		*	2,100
Financed by:			
Capital			
Cash introduced			2,000
Net profit for the year			800
			2,800
Drawings			(700)
Balance as at 31 Dec 20x7			2,100

* $2100 is the value of company in this moment.

[09] Equation

Final account = Trading & Profit & Loss A/C + Balance Sheet

[10] Function of Financial Statements

Financial Statements	Purpose
Trading, Profit & Loss a/c	Calculate the net profit
Balance Sheets	To show the assets and liabilities

[11] Examination

2003	S2		S4
2004	S2	S3	S4
2005	S2	S3	S4
2006	S2		

[12] Exercise

Please correct the format and figure

	$	$
Sales	10,000	
Less: Cost of Good Sold		
Opening Stock		1,000
Closing Stock	-500	
Purchase		200
Gross Profit	9,000	

Chapter 5

Bad Debts and Provision for Bad Debts

Content

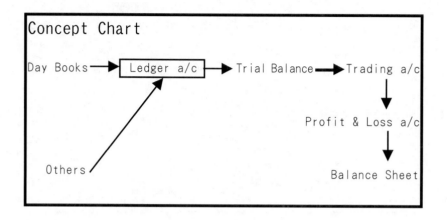

Concept Chart

Day Books ➝ Ledger a/c ➝ Trial Balance ➝ Trading a/c
↓
Profit & Loss a/c
↓
Others ↗
Balance Sheet

[01]   Introduction

Balance Sheet (extract)

Current Assets	$	$	$
Stock	X		
Debtors	X		

How to count it?

[02]   Bad Debt

When the firm finds that the debtor would not paid, that debt should be written off.

The accounting entries are:

Accounting entries	Explanation
Debit: Bad Debts account	To transfer the amount of uncollected debt to Bad Debts account
Credit: Debtors account	To reduce the amount of the debtor who is unable to paid

45

[03]   Why a business might create a provision for doubt debts

◎Prudence Concept – can avoid overstating profits and debtor
◎Matching Concept – can relate any loss for the period with the related sales in the same period

Classwork 1

We sold $1,000 goods to Mark on 1 January 20X7, but he became bankrupt. On 1 February 20X7 we sold $2,000 goods to Jimmy.   He managed to pay $200 on 17 May 20X7, but it became obvious that he would never be able to pay the final.

Mark	

Jimmy	

Bad Debt	

[04]   The posting for provision for bad debts

Provision for bad debts shows the $\boxed{\text{estimated}}$ amounts for debtors, who will be unable to pay.

Provision for Bad Debts			
	$		$
2001 Bal c/d	A	2001 Profit & loss a/c	X
2002 Profit & loss a/c	X	2002 Bal b/d	A
Bal c/d	B		
	X		X

Balance Sheets (Extracts)		
	$	$
2001 Debtors	X	
Less Provision for bad debts	A	X
2002 Debtors	X	
Less Provision for bad debts	B	X

[05]   Increasing the provision (Bal c/d > Bal b/d)

The double entry will be:

Debit : Profit and Loss Account

Credit : Provision for Bad Debts account

【Expenses】

47

[06]   Reducing the provision (Bal c/d < Bal b/d)

The double entry is:
> Debit : Provision for Bad Debts account
> Credit : Profit and Loss Account

【Income】

Classwork 2

Year	Credit Sales	Bad Debt	Net Debtor	%
1	10,000	100		5%
2	15,000	100		5%
3	20,000	100		1%

Debtor	

Bad Debt	

Profit & Loss account (extract)			
	$	$	$

Balance Sheet (extract)			
	$	$	$

[07] Different between bad debt and provision for bad debt

	Bad Debt	Provision for bad debt
Types	Expenses	Expenses / Incomes
Placing	PL / CA	PL / CA
Accuracy	Known with certainly	Estimate of possible losses

[08] Bad debts recovered

(a) Period: more than one year

First, reinstate the debt by making the following entries:

Debit: Bank account

Credit: Debtor

Debit: Debtor

Credit: Bad Debt Recover (incomes)

With the amount received.

(b) Period: on the current financial year

Dr: Bank

Cr: Bad Debt

Classwork 3

On 5 March 20x7, $2,500 was received from Mr So in settlement of the debt which had previously been written off as bad in 20x6.

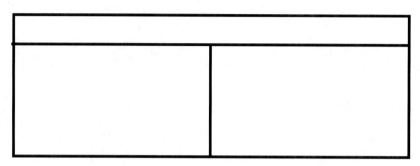

[09] Provision for Discounts

If it is desired to show the debtors' balance at a more realistic figure, another provision account namely "Provision for Discount Allowed" can be made.

Classwork 4

Debtor $100,000, provision for doubtful debts $2,000, discount for provision is 5%, bad debt $10,000

Balance Sheet (extract)			
	$	$	$
Debtors			
Less : Provision for doubtful debts			
Provision for discounts on debtor			

The firms find that the creditors' balance does not represent the amount which will paid. A Provision for Discount Received is made.

Dr: Provision for discount received

Cr: Profit & Loss a/c

[10] Examinations

2003	S3	
2004	S3	S4
2005		S4

Chapter 6

Depreciation & Disposal

Content

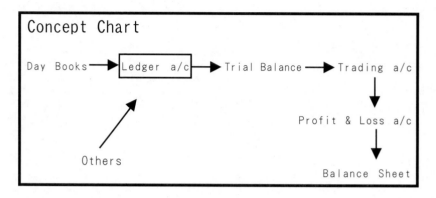

Concept Chart

Day Books ➝ Ledger a/c ➝ Trial Balance ➝ Trading a/c
                                              ↓
                         ↗   Profit & Loss a/c
                                              ↓
        Others   Balance Sheet

[01]   Introduction

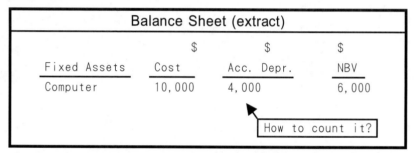

Balance Sheet (extract)

	$	$	$
	Cost	Acc. Depr.	NBV
Fixed Assets			
Computer	10,000	4,000	6,000

How to count it?

[02]   Depreciation of Fixed Assets

· Cost of the fixed asset consumed
· It is an expense for services consumed
· Matching concept

[03]   Causes of Depreciation

a. Physical depreciation

Wear & tear   Rust, rot and decay

b. Economics Factors

Obsolescence   Inadequacy

c. The time factor

d. Depletion

The wasting away of an asset as it is used as it is used up.

[04]   Land & Building

a. Land does not require a provision for depreciation (Except mines and quaries)

b. Buildings require a provision for depreciation

[05]   Methods of calculating depreciation charges

a. Straight Line Method – at an equal amount each year.

Classwork 1

If we bought a motor van for $20,000, thinking that we would keep if for 5 years and then sell it for $5,000.

Depreciation
= (Cost Price-Disposal Value)/No of estimated year of use

=

=

b. Reducing balance method – based on the net book value (NBV) of the asset brought forward from the previous year.

(Net Book Value=Cost – Accumulated Depreciation)

Classwork 2

If we bought a motor van for $20,000, thinking that we would keep if for 5 years and then will sell it for $5,000, depreciation rate is 20%.

Cost	20,000
First Year Depreciation ($20,000 x)	
Net book value at the end of the first year	___
Second Year Depreciation ($ x)	
Net book value at the end of the second year	___
Third Year Depreciation ($ x)	
Net book value at the end of the Third year	
	___

[06]   Double entry of Depreciation

The double entry is:

Dr: Profit & Loss a/c

Cr: Provision for depreciation a/c

Classwork 3

A machine is bought for $10,000 on 1 Jan 20X5. It is to be depreciated at the rate of 10% using the reducing balance method, prepare 2 years' account.

Machinery	

Provision for Depreciation	

[07]   Disposal

Upon the sale of a fixed asset, we will want to remove it from our ledger accounts.

Classwork 4

Use the above data, sell the machinery $3,500 at 2 Jan 20X7.

Disposal	

Classwork 5

Sell the machinery $9,000 at 2 Jan 20X7.

Disposal	

	Dr	Cr
Profit on Disposal		
Loss on Disposal		

[08]   Trade In (Additional Information)

賣資產五部曲：成、舊、錢、換、走	
1. Cost of FA 成 本	Dr: Disposal Cr: FA (Cost) (↓)
2. Accumulated Depreciation 累積折舊	Dr: Provision for Depreciation(↓) Cr: Disposal
3. Money Received 收 錢	Dr: Cash / Bank (↑) Cr: Disposal
4. Trade in value 以機換機	Dr: FA (↑) (trade in value) Cr: Disposal (same nature with cash receipt)
5. Profit or Loss 計完賺蝕就走	Profit Dr: Disposal Cr:Profit & Loss Loss Dr: Profit & Cr: Disposal

[09]   Examinations

2003	S2	S3	
2004	S2	S3	S4
2005	S2	S3	

Chapter 7

Cash Book

Content

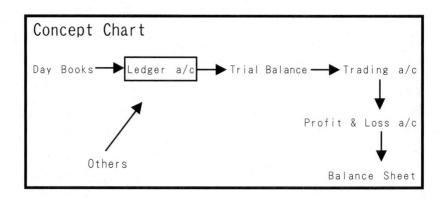

[01] Introduction

Cash Book = Bank a/c + Cash a/c + Discount Received + Discount Allowed

Dr → ↑   Cr → ↓

[02] Format

Cash Book

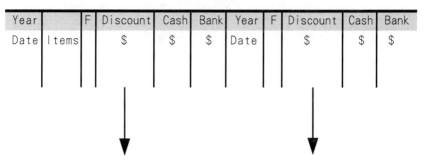

Year		F	Discount	Cash	Bank	Year	F	Discount	Cash	Bank
Date	Items		$	$	$	Date		$	$	$

Total of this column is transferred to Discounts Allowed a/c.

Total of this column is transferred to Discounts Received a/c.

◎The bank column contains details of the payment and receipts made by cheques

◎The cash column contains details of the payment and receipts made by cash

◎When the firm received the bank statement; check against the bank column in cash book

[03]   The use of folio columns

Folio columns used for entering reference numbers.

Classwork 1

An entry of receipt of a cheque from May whose account was on page 5 of the sales ledger, and the cheque recorded on page 7 of the cash book, would use the folio column as follows:

◎In the Cash Book : in the folio column would appear SL 5

◎In the Sales Ledger : in the folio column would appear CB 7

Year		F	Dis	Cash	Bank	Year	F	Dis	Cash	Bank
Date	Items		$	$	$	Date		$	$	$
Nov25	Sales									

Sales

20X7		F	$	20X7		F	$
				Nov 25			800

The advantages of folio columns

◎Speeds up reference to the other books

◎Make easier to find errors when a transaction has only been entered once.

[04] Cash Discount

Cash Discount is an allowance given to customers who pay early.

[05] Discounts Allowed & Discounts Received

It is usually expressed as a percentage, e.g 7%

Discounts Allowed	to customers when they pay early (expenses)
Discounts Received	from suppliers when the firm pay quickly. (revenue)

Alvin owned us $1,000. He paid us $950 and we allowed him $50 for the reward of quickly pay. We paid $980 for settle the debt of $1,000.

Cash Book

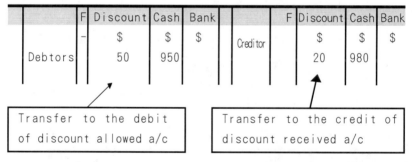

		F	Discount $	Cash $	Bank $		F	Discount $	Cash $	Bank $
Debtors		–	50	950		Creditor		20	980	

Transfer to the debit of discount allowed a/c

Transfer to the credit of discount received a/c

The discount columns are NOT part of the double entry.

[06] Cash paid into bank (cash decrease, bank increase)

20x7		F	Discount $	Cash $	Bank $	20x7		F	Discount $	Cash $	Bank $
Nov25	Cash	C			800	Nov25	Bank			800	

Contra is where both the debit and credit entries are shown in the cash book.

Contra sign "C" should be inserted in the "F" column

[07]   Balancing the Cash Book

To balance off an account:

Cash & Bank Column

1. same as the other ledger accounts
2. find out balance c/d

Discount Column

1. add up the amount

Classwork 2

Cash Book

		Discount	Cash	Bank			Discount	Cash	Bank
Cash	Bank								
		$	$	$			$	$	$
Jan 1	Sales		100	100	Jan 1	Creditors	30		200
	Debtors	20		980		Bal c/d			45
255									

[08]   Comparison of Trade Discount and Cash Discount 30

	Trade Discount	Cash Discount
Nature	A deduction from list price	A deduction of the sum to be paid
Purpose	To a trader buying goods in large quantities	To encourage customers to pay promptly
Double Entry	Only post in day books	Double entry system

[09]   Bank Overdrafts (Expenses > Income + Bal b/d)

The balance becomes a credit one. It is current liabilities (CL).

Classwork 3

Amy draw $2000 form her bank a/c of Hong Seng Bank by ATM.

Cash Book

20X7	F	D	C	B	20X7	F	D	C	B
May1					May1				

[10]   Examination

2004	S2	
2005	S2	S3

Chapter 8

Petty Cash Book

Content

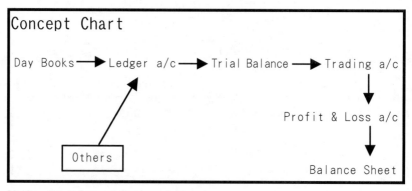

[01]　Division of the cash book

Petty cash book is a cash book for posting small cash payments.

[02]　Why we need it?

◎Many small cash payment to be made
◎Handling the small payments given to a junior book-keeper.

[03]　Voucher

Voucher is a form that shows what a payment. It included the following items:

◎Reference Number
◎Date
◎For what required
◎Amount
◎Signature
◎Passed by

Petty Cash Book

Receipts	Date	Detail	Voucher No.	Total	Postage	Clean- ing	Travel- ing	Sundries	Ledger Accounts
$				$	$	$	$	$	$

[04]   Imprest system

It is a system, which a refund is made of the total paid out in a period.

Classwork 1

20X0	The cashier gave the petty cashier	200
	The petty cashier paid out in 20x0	60
	Petty cash now in hand	140
	The cashier now gives the petty cashier	
	the amount spent	60
	Petty cash in hand at the end of 20x0	200
20X1	The petty cashier paid out in 20x1	56
	Petty cash now in hand	144
	The cashier now gives the petty cashier	
	the amount spent	____
	Petty cash in hand at the end of 20x1	=====

Classwork 2

Cathy Wong keeps her petty cash on the imprest system — amount being $1,000. The petty cash transactions for the month of March 20X7 were as follows:

Mar	
1	Petty cash in hand $800.
1	Petty cash restored to imprest amount
6	Bought pencil $60.
7	Paid wages $90.
20	Paid to Mr Lee, a creditor, $100.
21	Paid wages $70.
25	Bought envelopes $15.
28	Bought notepaper $20.

Cathy Wong

Petty Cash Book

Receipts	Date	Details	Total	Stationery	Wages	Postage	Ledger
$	20X7		$	$	$	$	$
800	Mar 1	Balance b/f					
600	1	Cash					
	6	Stationery	60	60			
	7	Wages	90		90		
	20	Mr. Lee	100				100
	21	Wages	70		70		
	25	Envelopes	15			15	
	28	Notepaper	20	20			
				80	160		100
	31	Balance c/d					
1,000			1,000				
	Apr 1	Balance b/d					
	1	Cash					

[05]   Imprest System in different period

1.Information

Mar 1   Received a cash float $600

Mar 2   Petrol $70

Mar 3   CD $60

Mar 4   Postage $90

Mar 5   A4 paper $60

2.Reimbursement at the end of a period

Petty Cash Book

Receipts	F	Date	Details	Voucher	Total	Motor Vehicles	Postage	Stationery
$				$	$	$	$	
600		1/3	Cash					
		2	Petro	I	70	70		
		3	CD	2	60			60
		4	Postage	3	90		90	
		5	Paper	4	60			60
					280	70	90	120
280		31	Cash					
			Bal c/d		600			
880					880			

3.Reimbursement at the beginning of a period

Petty Cash Book

Receipts	F	Date	Details	Voucher	Total	Motor Vehicles	Postage	Stationery
$				$	$	$	$	
600		1/3	Cash					
		2	Petro	I	70	70		
		3	CD	2	60			60
		4	Postage	3	90		90	
		5	Paper	4	60			60
					280	70	90	120
600		31	Bal c/d		320			
					600			
320								
280								

[06] Examinations

2003		S4
2004		
2005		S4
2006	S2	

Petty Cash Book

Chapter 9

Organisation

Content

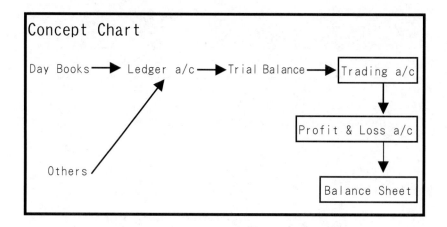

[01] Introduction

Club and society provide services for members, not for profit-making.

[02] Receipts and Payments Account

Receipts and Payments Account for the year ended

31 December 20X6

	$		$
Balance b/d	236	Groundsman's wages	728
Subscriptions received	1,148	Sports stadium expenses	296
Rent received	116	Committee expenses	58
		Printing and stationery	33
		Balance c/d	310
	1,500		1,500

Profit Making Firm	Non-profit Making Organization
1. Cash Book	1. Receipts and Payments Account

[03] Income and Expenditure Accounts

Income and expenditure account is an account for a non-profit making organisation to find the surplus or deficit made during a period.

Profit Making Firm	Non-profit Making Organization
1. Trading account	1. Trading a/c
2. profit and loss account	2. Income & Expenditure a/c
3. Net profit	3. Surplus
4. Net loss	4. Deficit

[04] Accumulated Fund

Accumulated Fund = Assets - Liabilities = Capital

Profit Making Firm	Non-profit Making Organization
1. Cash Book	1. Receipts and Payments Account
2. Drawing	2. Not available

Classwork 1

Sport Equipment $8,000, Club Premises $22,000, Furniture & Fittings $5,000, Rates & Insurances $200, Bank $2,000, Creditor $1,000, $ Bar Stock $2,000

Use the above information to prepare the statement of affairs.

Statement of Affairs

	$	$

[05] Drawing up Final Accounts

A separate trading account is to be prepared for a bar, where is made for one kind of income.

Classwork 2

Trial Balance as at 31 December 20X7

	$	$
Equipment	8,000	
Premises	29,000	
Subscriptions		6,000
Wages	4,000	
Furniture and fittings	5,000	
Rates and Rent	1,900	
Sundry expenses	600	
Accumulated fund 1 January 20X7		
Donations		300
Telephone ,postage & Printing	500	
Bank	2,000	
Bar purchases	9,500	
Creditors		1,000
Bar sales		14,000
Bar stocks 1 January 20X7	2,000	

Classwork 3

The following information is also available: (i) Bar stocks at 31 December 20X7 $2,362. (ii) Provision for depreciation: sports equipment $1,700; Furniture and fittings $1,315.

Bar Trading Account for the year ended 31 December 20X7

	$	$
Sales		
Less Cost of goods sold:		
Opening stock		
Purchases		
Less Closing stock	———	———
Bar profit to income and expenditure account		═══

Income and expenditure Account for the year ended 31 December 20X7

Income	$	$	$
Gross profit from bar			◄
Subscriptions			
Donations		———	
Less Expenditure			
Wages			
Rates and insurance			
Telephone and postage			
General expenses			
Provision for Depreciation:			
· Furniture and fittings			
· Sports equipment			
Surplus of income over expenditure	———	———	——— ═══

Balance Sheet as at 31 December 20x7

	$	$	$
Fixed assets	Cost	Acc. Depr.	NBV
Club premises			
Furniture and fittings			
Sports equipment			

Current Assets

Bar stocks

Cash in bank

Current Liabilities

Creditors for bar supplies

Net current assets

Financed by:

Accumulated fund

Balance at 1 January 20x7

Add Surplus

Balance as at 31 December 20x7

Surplus =   >

Deficit =   <

[06] Subscriptions

Subscriptions

20X7			$	20X7			$
Jan 1	In arrear b/d	(1)	360	Jan 1	In advance b/d	(2)	80
Dec 31	I & E (Bal fig)		*7,220	Dec 31	Bank	(4)	7,420
	In advance c/d	(3)	140		In arrear c/d	(5)	220
			7,720				
20X8				20X8			
Jan 1	In arrear b/d	(5)	220	Jan 1	In advance b/d	(3)	140

[07] Donations

Any donations received are shown as income in the year that they are received.

[08] Entrance Fees

New members often need to pay an entrance fee when they join. Entrance fees are normally included as income in the year that they are received. Sometimes it would treat as accumulated fund at examination.

[09] The double entry of donations and entrance fees

Description	Double Entry
Donation received or entrance fee received	Dr: Cash Cr: Donation / Entrance Fee Received

[10]   Life Membership

Sometimes members can pay one amount for membership, and then this membership will last for their lifetime.

[11]   Different between I & E and R & P

	Receipts & payment a/c	Income & Expenditure a/c
Nature	Record the cash in and out in the period briefly	Calculate the surplus and deficit

[12]   Examinations

2003			
2004	S2		
2005			

Chapter 10

Bank Reconciliation Statement

Content

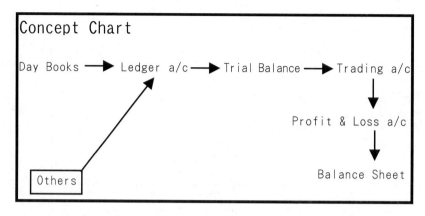

[01] The purpose of the bank reconciliation statement

◎ explain the reasons for the differences
◎ to identify errors and omissions in bank account and bank statement.

[02] Reasons for differences

Due to the timing difference, omissions and errors made by the bank or the firm itself. The balances of the bank statement by bank and the bank account by firm did not agree.

Sources	Items
Internal	Uncredited Cheque
	Error on Cash Book
External	Credit Transfer
	Standing Order
	Bank Interest
	Bank Charge
	Bank Lodgment
	Error on Bank Statement
Third Parties	Unpresented Cheque
	Dishonoured Cheque

(i)   Dishonoured cheques

They are cheques deposited but subsequently returned by the bank due to the failure of the drawer to pay.

Example

Jan 1   A debtor, Peter give us a cheque $100

Jan 2   Bank tell us the Peter's cheque dishonourable

Cash Book		
$	[2]	$
Jan 1 Debtor — Peter[1]	Jan 2 Debtor — Peter	
100	(Dishonourable)	100

Classwork 1

Feb 1   Peter give us $1,000 cheque and $200 cash

Feb 5   It has dishonorable for Peter

Cash Book	

(ii) Bank charges

They are charges made by the bank to the company for banking services provided.

(iii) Standing orders S/O

The instructions from the firm to the bank to make regular payments such as rent.

(iv) Direct debits D/D

The instructions from the firm to the bank to make non-regular payments such as electricity.

(v) Credit transfers C/T

They are collections from customers directly through the bank.

(vi) Interest allowed by the bank

They are interest received for fixed or time deposits.

(vii) Uncredited items

The deposits occurred too close to the cut-off date of the bank statement and so do not appear on the statement such as 31 Dec. They will appear on the next statement.

(viii) Unpresented cheques

They are cheques issued by the firm that have not yet been presented. The firm has not paid any.

[03]   Nature of the cash book and bank statement

Cash Book = Asset

Cash Book

+	$	−	$
A debit represents an increase		A credit represents a decrease	

81

The balance as per the bank statement is a liability to the bank, so:

	Bank statement		
	Dr	Cr	Balance
	$	$	$
	Withdrawal	Deposit	Represents the amount owed to the film
	—	+	

[04] Prepare the bank reconciliation statement

First Step (Updated Cash Book)			
	$		$
Bal b/d	X	Dishonoured cheques	X
Credit transfers	X	Bank charges	X
Bank interest received	X	Standing orders	X
Wrong credit by the bank	X	Direct Debit	X
		Wrong debit by the bank	X
		Balance c/d	A
	X		X

Second Step (Bank reconciliation statement as at XXX)	
	$
Balance as per cash book	A
Add: Unpresented cheques	X
	X
Less : Uncredited items	X
	X
Balance as per bank statement	X

Classwork 2

Cash Book

20x7		$	20x7		$
Jun1	Balance b/f	2,000	Jun5	D. Wong	150
7	B. Moon	160	12	J. Yau #1004	400
11	Mr Ng 200	200	17	Peter #1005	100

Bank Statement

20x7		Dr.$	Cr.$	Balance $
Jun 1	Balance b/f			2000
7	Cheque		160	2,160
8	Wong	150		2,010
9	Bank Charge	100		1,910
10	Mr So		1,000	2,910

You are required to:

(a) Write the cash book up-to-date to take the above into account.

(b) Draw up a bank reconciliation statement as on 30 June 20x7 as begin as update cash book balance.

(c) Draw up a bank reconciliation statement as on 30 June 20x7 as begin as bank statement balance.

[05]   Bank Overdrafts

The process is the same as those needed when the account is not overdrawn.

Classwork 3

Cash Book

Bal b/d	100	Rent S/O	1000
May C/T	50	Electricity D/D	100
Bal c/d		Bank Charge	20

Bank Reconciliation Statement

	$	$
Balance as per Cash Book		

[06]   Dishonoured Cheque

After banked the cheque, the customer's bank will not pay the amount due on the cheque, the customer's bank has failed to "honour' the cheque.

Classwork 4

In 1 Jan 20x7, received $1,000 from May.

In 3 Jan 20x7, our bank tell us May's account has not sufficient funds.

Bank

Debtor-May

[07]   Examination

2003	S2	
2004		S3
2005	S2	S3
2006	S2	

Chapter 11

Journal

Content

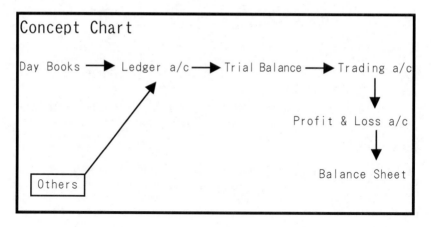

[01] The Nature of Journal

Journal is used to record those transactions, that cannot go the subsidiary day books, i.e. Cash Book, Sales Journal, Purchase Journal, Return Inwards Journal, Return Outwards Journal, and Petty Cash Book. It is NOT part of the Double Entry System.

Journal				Page
Date	Particulars	F	Dr	Cr
			$	$

[02] Transaction to be recorded in Journal

◎ Opening entries

◎ Purchase and sale of fixed assets on credit

◎ The correction of errors

◎ All other transfer that cannot record in any day books

Classwork 1

Mr So invested Premises $350,000, Motor Van $60,000 and Cash $28,000 to start his business on January 1, 20x7. To begin writing the books, these items must be entered into the Journal.

Journal			P.1
Date 20x7 Jan 1	Particulars	F	Dr $ Cr $
	Being Assets and Capital at opening new set of books		

Classwork 2

Maria Company after being in business for a number of years, decided to keep a proper set of double entry books. On January 1, 20x7, its assets and liabilities were as follows:

◎ Assets :

Fixtures and Fittings $20,000 , Office Equipment $10,000 , Motor Vehicle $26,000 , Debtors $17,000 , Cash $10,000

◎ Liabilities:

Creditors $21,000

Journal			P.1
Date Particulars	F	Dr	Cr
20x7		$	$
Jan 1			
Being Assets, Liabilities and Capital at opening new set of books			

Classwork 3

A typewriter was bought on credit from Kowloon Co. for $4,000 on January 3, 20x7.

Journal			P.1
Date Particulars	F	Dr	Cr
20x7		$	$
Jan 3			
Purchase of typewriter on credit.			

Classwork 4

On April 16, 20x7, an office desk, bought for use in the firm for $1,500, was sold on credit to Kennedy Co. for $1,300 and no provision for depreciation had been made.

Journal				P.3
Date	Particulars	F	Dr	Cr
20x7			$	$
Apr 16				
	Sale of office desk on credit			
	Sold office equipment to Kennedy Co. on credit			
	Loss on sales of office equipment			

Classwork 5

Stock $500 was taken out as sample for promotion on May 16, 20x7, $100 for promotion and the balance for private use.

Journal				P.7
Date	Particulars	F	Dr	Cr
			$	$
20x7				
	Stock withdrew as sample for promotion			

[03] Errors Affecting Trial Balance Agreement

◎ forgetting to post either the sides.
◎ wrong calculation in the day book.
◎ wrong calculation the balance of an account
◎ Debit and credit amounts are different
◎ Post an entry into the wrong side of an account.

[04] The Effect of Errors on the Trial Balance

a)   Total Dr bal **>** Total Cr bal

	Dr	Cr
Case 1	Overstate	Correct
Case 2	Correct	Understate
Case 3	Incorrect	Incorrect

b)   Total Cr bal **>** Total Dr bal

	Dr	Cr
Case 1	Understate	Correct
Case 2	Correct	Overstate
Case 3	Incorrect	Incorrect

[05] Effect of Errors on Profit

If errors are made in the Revenue Account, Expenses Account and Stock, the amount of gross profit and net profit will be affected.

Account	Error	Effect on	
		Gross Profit	Net Profit
Stock (Opening)	Understate	Overstate	Overstate
	Overstate	Understate	Understate
Stock (Closing)	Understate	Overstate	Overstate
	Overstate	Understate	Understate
Purchase	Understate	Understate	Understate
	Overstate	Overstate	Overstate
Purchase Returns	Understate	Understate	Understate
	Overstate	Overstate	Overstate
Sales	Understate	Overstate	Overstate
	Overstate	Understate	Understate
Sales Return	Understate	Understate	Understate
	Overstate	Overstate	Overstate
Trading Expense	Understate	Overstate	Overstate
	Overstate	Understate	Understate
Other Expenses	Understate	No Effect	Overstate
	Overstate		Understate
Other Income	Understate	No Effect	Understate
	Overstate		Overstate

Classwork 6

Office Furniture had been purchased for $3,600 on credit from Kowloon Co. on Jan 17, 20x6 that had not been recorded. The bookkeeper found this error on April 11, 20x7.

Journal					P.16
Date	Particulars		F	Dr	Cr
20x7				$	$
Apr 11					
	Office Furniture purchased from				
	Kowloon Co. on Jan 17, 20x7				
	had been omitted				

Classwork 7

$20,000 of goods sold on credit to Mr Chan had been entered in the account of Mr Chen. The Bookkeeper found this error on May 1, 20x7.

Journal					P.16
Date	Particulars		F	Dr	Cr
20x7				$	$
May 1					
	Sales entered in wrong cus-				
	tomer account				

Classwork 8

A sale of goods to Aberdeen Co. $690 had been entered in the books as $960 on Apr 11, 20x7. The Bookkeeper found this error on June, 20x7.

Date	Particulars	F	Dr	Cr
	Journal			**P.17**
20x7			$	$
Jun 7				
	Credit sales had been recorded			
	overstating, now correct.			

Classwork 9

$20,000 incurred to repair the Motor Vehicles in order to extent the life of that Motor Vehicle had been debited to Repair Expenses Account on Mar 19, 20x7. The bookkeeper found this error on July 26, 20x7.

Date	Particulars	F	Dr	Cr
	Journal			**P.19**
20x7			$	$
Jul 26				
	Cost of Motor Vehicles had			
	been wrongly debited to Re-			
	pair Expenses			

Classwork 10

Cheque $3,000 received from Kwun Tong Co. had been entered twice in the Cash Book on Apr 16, 20x7. the bookkeeper found this error on Aug 18, 20x7.

Journal					P.16
Date	Particulars		F	Dr	Cr
20x7	Correcting Entries			$	$
Aug 18					
	Cheque received had been entered twice in the Cash Book				

Classwork 11

Selling Expenses Account was overstated by $150 and the Sales Account was overstated by $150 too. The bookkeeper found these errors on September 16, 20x7.

Journal					P.23
Date	Particulars		F	Dr	Cr
20x7	Correcting Entries			$	$
Sept 16					
	Both the Selling and Telephone Expenses Account had been overstated by $100				

Classwork 12

Purchase of goods for cash $4,600 had been entered in the debit side of Cash account and posted to the Credit side of the Purchases Account. Error was found on 11 Nov 20x7.

Journal					P.27
Date	Particulars		F	Dr	Cr
20x7				$	$
Nov 11					
	Purchased goods for cash				
	had been recorded in reverse				

[06] Examinations

2003
2004
2005

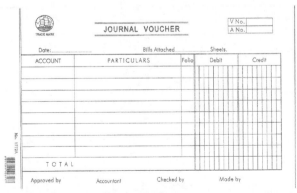

Journal Voucher

Chapter 12

Adjustment for Final Account

Content

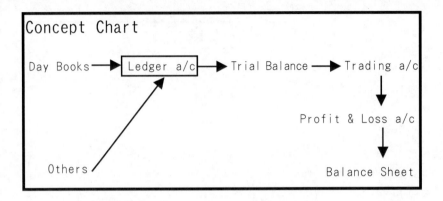

[01] Introduction

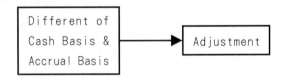

Adjustments:

◎ Depreciation Expenses and Provision for Depreciation

◎ Bad Debts and Provision for Bad Debts

◎ Accrued Expenses

◎ Prepaid Expenses

◎ Accrued Revenue

◎ Prepaid Revenue

[02] Expenses

Expenses

	$		$
Prepay b/d	X	Accrual b/d	X
Bank	X	Profit & Loss (Bal Fig)	X
Accrual c/d	X	Prepay b/d	X
	X		X

	Profit & Loss account	Balance Sheet
Expenses unpaid	Add	Current Liability
Expense prepaid	Less	Current Assets

Classwork 1

	31/12/20X6	31/12/20X7
Prepay Rent	$100	$200
Accrual Rent	$200	$400
Paid for Rent	$1,000	$5,000

Rent

[02] Income

Income			
	$		$
In arrear b/d	X	In advance b/d	X
Profit & Loss (Bal Fig)	X	Bank	X
In advance c/d	X	In arrear c/d	X
	X		X

	Profit & Loss account	Balance Sheet
Income in arrear	Add	Current Asset
Income in advance	Less	Current Liability

Classwork 2

	31/12/20x6	31/12/20x7
Subscription in arrear	$1,000	$1,500
Subscription in advance	$200	$400
Bank	$5,000	$7,000

[03] Balance Sheet

Balance Sheet (extract)			
Current Assets	$	$	$
Stock	100		
Debtors	200		
Prepayment			
Subscription in arrear			
Bank	1,000		
Less: Amount due within one year			
Creditor	150		
Accrual Expenses			
Subscription in advance			
Net Current Assets			

[04] Adjustment on final account

The following balances were part of the trial balance of Maria on 31 December 20X7 :

	Dr	Cr
	$	$
Stock at 1 January 20x6	2,070	
Sales		22,000
Purchase	11,180	
Rent	650	
Wages and salaries	2,160	
Insurance	590	
Bad debt	270	
Telephone	300	
General expenses	180	

On 31 December 20X7 you ascertain that :
(a) The rent for four months of 2007 $160 had prepaid in 20x6.
(b) $290 owed for wages and salaries.
(c) Insurance had been prepaid $190.
(d) A telephone bill of $110 was owed.
(e) Stock on 31 December 20x7 was valued at $3,910.

Draw up Maria's trading and profit and loss account for the year ended 31 December 20X7 using the vertical style.

STEP

(a) It is prepayment; less on the Trial Balance
(b) It is accrual expense; add on the trial balance
(c) It is prepayment.
(d) It is accrual
(e) Closing stock

[05]   Examination

2003	S2	
2004	S2	S3
2005	S2	

Chapter 13

Capital and Revenue Expenditure

Content

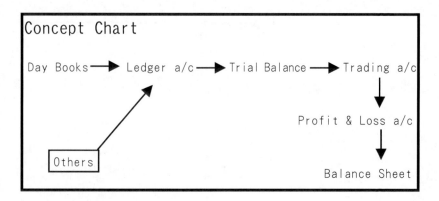

[01]   Introduction

Expenditure	Details
Capital Expenditure	For buy Fixed Assets For the use of new Fixed Assets Upgrade the value of existing Fixed Assets
Revenue Expenditure	For business operation Cannot upgrade the value of existing Fixed Assets
Joint Expenditure	Mixed with the above expenditures

[02]   Revenue

a.Revenue expenditure

Revenue expenditure is incurred in keeping all fixed assets within the accounting period.

b.Revenue receipts

Revenue receipts are the earnings from the operations of the business. Example are all goods sold and interest on investments.

[03] Capital

Capital expenditure means the purchase of fixed assets or expenditure which adds to the value of existing fixed assets. Examples are adding fitment to equipment.

Example 1

Expenditure	Type of Expenditure
a. Purchase of a motor van	Capital
b. Petrol costs for motor van	Revenue
c. Repairs to motor van	Revenue
d. Painting outside of new building	Capital
e. Five years later − repainting outside of building in (d)	Revenue
f. Extension to factory premises	Capital
g. Goods bought for resale	Revenue
h. Cleaning of office	Revenue
i. Wages and salaries $200,000 of which $20,000 paid to employees engaged on building an addition to the office premises	Capital $20,000 Revenue $180,000

[04] Compare

Capital Expenditure	Revenue Expenditure
It has long term nature	It has short term nature
It has an enduring influence on the profit-making capacity of the business	It is not a temporary influence on the profit-making capacity of the business
It is concerned with cost of purchasing fixed assets or adding to the value of an existing assets In addition, it also included other costs necessary to get the fixed asset to operate	It is not concerned with adding to the value of fixed assets, but represents the costs of running the business on day to day basis It is chargeable to the Trading, Profit & Loss account
It will result in increased figures for fixed assets in the Balance Sheet	

[04] Why it is so important to know their different?

The revenue expenditure would affect the profits. Overstate the revenue expenditure → net profit decrease.

[05] Examination

2003		
2004		S4
2005		
2006	S2	

106

Chapter 14

Correction of Error 1

Content

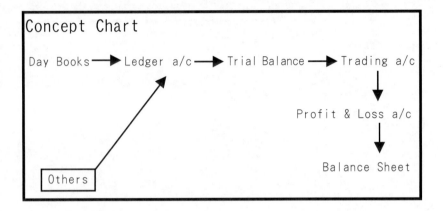

[01]   Introduction

Errors in the books have:

1. do not show in Balance Sheet
2. show in the Trial Balance

When book-keepers find the mistakes, they must correct them by journal.

[02]   Errors of Omission

The absence of both ledger accounts.

Example 1

Office furniture purchased for $3,000 on credit from So Limited and has not been recorded.

	Dr	Cr
Right	Office Furniture	So Ltd
Actually	———	———

Correction

Journal		
	Dr	Cr
	$	$
Office furniture		
So Ltd		
Correction of omission of capital purchase.		

[03] Errors of Commission

The wrong posting which is the same class.

Example 2

A sale of goods $1,200 on credit to T So has been entered in the account of D So.

	Dr	Cr
Right	Debtor - T So	Sales
Actually	Debtor - D So	Sales

Journal		
	Dr	Cr
	$	$
Debtor - T So	1,275	
Debtor - D So		1,275
Sales entered in wrong customer account, now corrected.		

[04]   Errors of Principle

The wrong posting which is the ⃞different⃞ class.

Example 3

The payment of $1,000 for repairs has been wrongly post to motor van account.

	Dr	Cr
Right	Repairs	
Actually		

Journal		
	Dr $	Cr $
Repairs		
Motor vehicles		
Repairs wrongly debited to motor vehicle account, now corrected.		

[05]   Errors of Original Entry

The original figures are posted incorrectly in both accounts.

Example 4

Printing account is over-stated by $20 and sales account also is over-stated by $20.

	Dr	Cr
Right	Printing $100	Sales $1000
Actually		

Correction

Journal		
	Dr $	Cr $
Sales		
Over-statement in each of sales account and print-ing account, now corrected.		

[06]   Errors of Compensations

Errors which posted in different sides in two accounts

Example 5

A sale of goods $690 on credit to Lee has been entered in the books as $960.

	Dr	Cr
Right		
Actually		

Journal		
	Dr $	Cr $
Sales understated, now corrected.		

[06] Reversal of Entries

The account, which should be debit, has been credit, vice versa.

Example 6

Purchase of goods by cheque $1,000 has been entered wrong in both debit and credit sides.

	Dr	Cr
Right		
Actually		

Purchases			
	$		$
Bank		Bal b/d	1,000

Bank			
	$		$
Bal b/d	10,000	Purchases	

[07] Examinations

2003		S3	
2004			
2005			
2006	S2		

Chapter 15

Correction of Error 2

Content

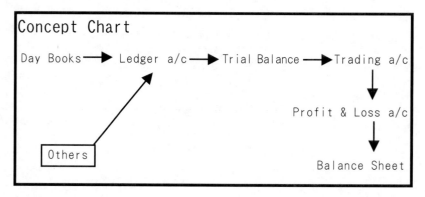

Concept Chart

Day Books → Ledger a/c → Trial Balance → Trading a/c
                                                ↓
                                        Profit & Loss a/c
                                                ↓
Others ↗   Balance Sheet

[01]   Errors of Extraction

Example 1

Omission in the extraction of Sundry expenses $1,000

Correction :	Dr	Sundry expenses	$1,000
	Cr	Suspense	$1,000

[02]   Errors in Posting

Advertising expenses $420 posted as $240

Example 2

Correction :	Dr	Advertising ($420 - $240)	$180
	Cr	Suspense	$180

[03]   Errors of Incomplete Double Entry

The posting of cash payment for wages $700 has been omitted

Example 3

Correction :	Dr	Wages	$700
	Cr	Suspense	$700

[04]   Errors in Addition

The total of the Returns inward books has been over-cast by $600

Example 4

Correction :	Dr	Suspense	$600
	Cr	Returns inwards	$600

[05]   Errors in the Balancing of Accounts

Example 5

The balance of the Cash account has been under-stated to the extent of $500

Correction :	Dr	Cash	$500
	Cr	Suspense	$500

[06]   Statement of corrected Net Profit

When we correct the errors, the net profit or loss will have been incorrectly. We also need to adjust the net profit or loss by the statement.

Statement of adjusted the net profit		
	$	$
Net Profit before adjusted		100,000
Add: (4) Overcast of Return Inwards		
Less: (1) Omitted of Sundry Expense		
(2)Error posting on Advertising		
(3) Wage omitted		
Net Profit after adjusted		

[07]   Revised Balance Sheet

As adjusting the entries have been made, correct the net profit for the period, the balance sheet should be revised to reflect the actual financial status of the business.

[08]   Examination

2003		S4
2004	S2	
2005		

Chapter 16

Stock Valuation

[01]　Introduction

If you buy a lot of good and some of them is not so famous. what would you do? You can sell it in a great discount.

Another case, after the "black rain", some of the goods have destroyed. You must spend some expenses before you sell it.

In the above two case, what should we do in book-keeping?

[02]　Lower of cost or market value

Stock must be valued at the lower of its cost or net realizable value.

Saleable value — Expenses needed before completion of sale = Net realizable value (NRV)

Classwork 1

If the stock at cost valued $500 but the net realizable value is $400. Then how much would you be used as the value in the Trading a/c and Balance Sheet?

Classwork 2

Stock Item　　　　:PC01
Number in stock :100
Cost price　　　　:£150
Selling price　　:Was £300 but reduced to £250 in February
　　　　　　　　　　 20X7

Stock Item　　　　:PC02
Number in stock 15
Cost price　　　　:£165
Selling price　　:£200 but 5 refrigerators required repairs
　　　　　　　　　　 costing £60 each
Stock Item　　　　:PC03

	Cost	NRV
PC01		
PC02		
PC03		

[03] Margin and Markup

Since the closing stock are not valued at the selling price, adjustments for goods sold and sale returns should therefore be converted into cost by deducting the profit loading from their sales values. Markup and margin are terms used to denote the profit of goods sold.

◎Markup = GP / COFS

◎Margin = GP / Selling Price

Classwork 3

If cost price is $400,000; gross profit is $100,000, then what is his markup?

Classwork 4

If cost price is $400,000; gross profit is $100,000, then what is his margin?

Level 2

《LCCI Level 2 中級會計精讀備試天書》

Content 目録 ////

本書附送 LCCI 增值錦囊，請到以下網址下載：

http://www.systech.hk/lcci-note.rar

[01] Introduction

Balance Sheet (extract)

Fixed Assets	$ Cost	$ Acc. Depr.	$ NBV
Office Equipment	10,000	4,000	6,000

How to count?

Entities often buy assets intended for long-term use in the business that will provide a benefit for profit over a number of accounting periods. Such assets are known as Fixed Assets.

Under the matching concept, it is necessary to charge a part of the cost of a fixed asset to EACH accounting period that is expected to benefit from its use. Such a charge is known as Depreciation.

所有 classwork 的工作紙儲存於 CD 內，可以預先列印。

[02] Methods of calculating depreciation charges

a. *Straight Line Method - at an equal amount each year. (simply)*

Depreciation per annum ← expense in P & L	=	Cost Price - Disposal Value / No. of estimated year of use

Classwork 1

If we bought a motor van for $20,000, thinking that we would keep it for 5 years and then sell it for $5,000.

[工作紙可於 CD 內下載]

b. *Reducing balance method – based on the net book value of the asset*

Depreciation Expense	=	(Cost – Accumulated Depreciation) x %

Classwork 2

If we bought a motor van for $20,000, thinking that we would keep it for 5 years and then will sell it for $5,000, depreciation rate is 20%.

c. *Revalued Method* 重估方法

| Bal b/d | - | Depreciation Expense | - | Disposal | = | Revalued Value |

Classwork 3
Beginning of the period $120,000   End of the period $100,000
Buy the new one $50,000   Cost of old one $40,000

---

d) *Sum-of-the-year-digits*

Depreciation is apportioned according to the digit assigned in each financial year.

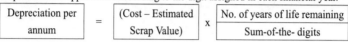

| Depreciation per annum | = | (Cost – Estimated Scrap Value) | x | No. of years of life remaining / Sum-of-the- digits |

Classwork 4
The original cost of a equipment is $50,000.   The expected useful life and the estimated scrap value are 5 years and $5,000 respectively.

---

e) *Units of production / Machinery hours method*

Depreciation is calculated on the number of units of production in each financial period out of the total available units.

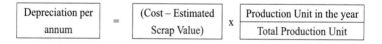

| Depreciation per annum | = | (Cost – Estimated Scrap Value) | x | Production Unit in the year / Total Production Unit |

Classwork 5
The original cost of a equipment is $100,000.   The estimated scrap value and the estimated number of production are $10,000 and 5,000 units respectively.

---

[03] Double entry of Provision for Depreciation

We usually used to maintain each fixed assets at its cost while hold another ledger account for depreciation.   This account is called "provision for depreciation".

Which method of depreciation should be used?

A different method might use for a certain asset.   For example, it is a fact that motor vehicles lose value heavily early in their life and less so in later years.   The reducing balance method therefore use appropriate because it indicates the way in which the value of the asset is lost.

Dr: Profit & Loss a/c   Cr: Provision for depreciation a/c

> *Classwork 6*
> A machine is bought for $10,000 on 1 Jan 20X5. It is to be depreciated at the rate of
> 10% using the reducing balance method, prepare 2 years' account.

The fixed assets will be fully depreciated some years later. If the fixed assets would
still be operated after then, no depreciation would be calculated in following years and
the net book value would be shown as "0".

[04] Disposal

Upon the sale of a fixed asset, we will want to remove it from our ledger accounts.
Some items which must appear in the disposal of fixed assets a/c are as follows:

the original value of the asset at cost	+	the accumulated depreciation up to the date of sale	+	the sale prices of the asset.

	DEBIT (Dr)	CREDIT (Cr)
[1]	Disposal of fixed asset account	Fixed asset a/c with the <u>cost</u> price.
[2]	Provision for depreciation account	Disposal of fixed asset account with the accumulated depreciation value
[3]	Cash book	Disposal of fixed asset account with the sale price of the asset.
[4]	The balance on the disposal account is the profit or loss on disposal and the corresponding double entry is recorded in the profit and loss account.	

> *Classwork 7:* Use the above data, sell the machinery $3,500 at 2 Jan 20X8

> *Classwork 8:* Sell the machinery $9,000 at 2 Jan 20X8

	Dr	Cr
Profit on Disposal	Disposal	Profit & Loss
Loss on Disposal	Profit or Loss	Disposal

Classwork 9

Richard Co. sold the machine at $2,000, which has bought at $10,000 and fully depreciation.　　The depreciation rate is 20% by reducing balance method.

折舊是一課必考的課題，可以獨立一題或在 Final Accounts，所以考生必須熟讀計算方法及 Disposal 的格式。

= 本章完　End of Chapter =

Chapter 2
Provisions for doubtful debts & discounts

[01] Introduction

In this chapter you will learn the situation where there is some doubtful about the payment of the debt.

所有 classwork 的工作紙儲存於 CD 內，可以預先列印。

[02] Why a business might create a provision for doubtful debts

A provision for any bad and doubtful debt may be made when a firm thinks that there would be problems in recovering a debt.

- Prudence Concept - can avoid overstating profits and debtor
- Matching Concept – can relate any loss for the period with the related sales in the same period

Provision for doubtful debts shows the estimated amounts for debtors, who will be unable to pay their accounts, at the balance sheet date.

Using current period outstanding debtors' balance, a company would forecast the new provision of doubtful debts for the coming financial period.

Provision for doubtful debts

= | Debtors after bad debts (net debtors) | x | rate (%) |

[03] The posting for provision for doubtful debts

The provision reduces the value of debtors in an attempt to give their real worth to the business.   It is a contra-asset item charged as an expense.

The double entry will be: (10 →12)

Dr	Cr
Profit and Loss Account	Provision for Bad Debts account

In Year 1, The amount of debtor is $100,000 and the rate of provision for doubtful debt is 10% of outstanding debts.   In Year 2, the amount of debtor is $120,000 and the rate is remaining unchanged.

Provision for Doubtful Debts

Year 2		$	Year 2		$
			Jan 1	Bal b/d [1]	10,000
				(100,000 x 10%)	
Dec 31	Bal c/d [2]	12,000	Dec 31	Profit & Loss (bal)	2,000
	(120,000 x 10%)				
		12,000			12,000
			Year 3		
			Jan 1	Bal b/d	12,000

> **Classwork 1**
>
> In Dec 20X7, our provision for doubtful debt increases from $10,000 to $12,000.

Should the amount required for the provision have to be reduced, the provision account must reduce its credit balance.   The corresponding credit entry will be made in the profit and loss account as <u>income</u>.

The double entry is: (12→10)

Dr	Cr
Provision for Bad Debts account	Profit & Loss account

In Year 2, The amount of debtor is $120,000 and the rate of provision for doubtful debt is 10% of outstanding debts.   In Year 3, the amount of debtor is $110,000 and the rate is remaining unchanged.

Provision for Doubtful Debts

Year 2		$	Year 2		$
			Jan 1	Bal b/d	10,000
	(120,000 x 10%)			(100,000 x 10%)	
Dec 31	Bal c/d	12,000	Dec 31	Profit & Loss (bal)	2,000
		12,000			12,000
Year 3			Year 3		
Dec 31	Profit & Loss (bal)	1,000	Jan 1	Bal b/d [1]	12,000
	Bal c/d [2] (110,000 x 10%)	11,000			
		12,000			12,000
			Year 4		
			Jan 1	Bal b/d	11,000

In Dec 20X7, our provision for doubtful debt decreases from $12,000 to $9,000.

Provision for Bad Debts

		$			$
2001	Bal c/d	X	2001	Profit & loss a/c	X
2002	Profit &n loss a/c	X	2002	Bal b/d	X
	Bal c/d	X			
		X			X

Profit and Loss Account (extracts)

		$			$
2001	Provision for Bad debts (increase)	X		(reduction)	
			2002	Provision for Bad Debts	X

Balance Sheets (Extracts)

Current Assets		$	$
2001	Debtors	X	
	Less Provision for bad debts	X	X
2002	Debtors	X	
	Less Provision for bad debts	X	X

Classwork 3

Year	Debtor Balance ($)	Bad Debt ($)	Net Debtor ($)	Doubtful debt rate
1	10,000	100		5%
2	15,000	100		5%
3	20,000	100		1%

Prepare the profit & loss account and balance sheet.

[04] Different between bad debt and provision for doubtful debt

	Bad Debt	Provision for bad debt
Types	Expenses	Expenses / Incomes
LCCI	Level 1	Level 2 & 3
Placing	PL	PL / CA
Accuracy	Known with certainly	Estimate of possible losses

[05] Provision for Discounts

In the business environment, it is estimation of the discounts allowed which occurred in coming period, based on the existing level of debtors balance. The amount of the Provision for Discounts Allowed is calculated as a certain percentage of net debtor balance which after the deduction provision for bad and doubtful debts at the end of each financial period.

Classwork 4

Description	2005 ($)	2006 ($)	2007 ($)
Debtors bal c/d	2,000	3,000	2,000
Bad Debts	500	400	500
Net Debtors	1,500	2,600	1,500
Provision for Bad Debts (%)	5	10	15
Provision for Bad Debts ($)	(75)	(260)	(225)
Provision for discount allowed (%)	4	3	4
Provision for discount allowed ($)	(57)	(70.2)	(51)

Classwork 5

Debtor $100,000, provision for doubtful debts $2,000, discount for provision is 5%, bad debt $10,000. Prepare the balance sheet.

以往 Provision for doubtful debt 是屬於 Level 1 的課題,現改為 Level 2,本人覺得其中一個可能性是因為很多 Level 1 的考生未能掌握其中要點,尤其是 double entry 的過程,考生必須熟習其中過程,因為此題目是經常混入 Balance Sheet 內的,是屬於大熱的課題。

= 本章完 End of Chapter =

Chapter 3
Income & Expenditure a/c

[01] Introduction

Club and society provides services for member. It is not for profit-making.

所有 classwork 的工作紙儲存於 CD 內，可以預先列印。

[02] Types of Organisations

Private Enterprise (profit making)

Sole Proprietor + Partnership + Limited Company + Franchising + Co-operative

Public Enterprise (Non-profit making)

Public Corporations + Government Departments + Local Authorities

Non-profit Making Organisations:

Club, Associations, Parties, Hospitals, Schools, Charitable bodies

[03] Receipts and Payments Account

Profit Making Firm	Non-profit Making Organization
Cash Book	Receipts and Payments Account

Receipts and Payments a/c

	$		$
Bal b/d	200	Bar wages	700
Subscriptions	1000	Sports stadium expenses	200
Rent received	100	Bal c/d	400
	1300		1300

[04] Income and Expenditure Accounts

Income and expenditure account is an account for a non-profit making organization to find the surplus or deficit made during a period.

Profit Making Firm	Non-profit Making Organization
1. Trading account	1. Trading account
● sales	● Bar taking
● gross profit / loss	● Bar profit / loss
2. Profit and Loss account	2. Income & Expenditure a/c
3. Net profit	3. Surplus
4. Net loss	4. Deficit

[05] Accumulated Fund

Accumulated Fund = Assets - Liabilities

Profit Making Firm	Non-profit Making Organization
1. Capital	1. Accumulated Fund
2. Drawing	2. Not available

Classwork 1

Sport Equipment $8,000, Club Premises $22,000, Furniture & Fittings $5,000, Prepaid Rates $200, Bank $2,000, Creditor $1,000, $ Bar Stock $2,000
Use the above information to prepare the statement of affairs.

---

[06] Subscriptions (Income)

The major source of incomes of a non-profit making organisation is the subscriptions of members.

Subscriptions

20X7			$	20X7			$
Jan 1	In arrear b/d	(1)	300	Jan 1	In advance b/d	(2)	100
Dec 31	I & E (Bal fig)		*7000	Dec 31	Bank	(3)	7,000
	In advance c/d	(3)	100		In arrear c/d	(5)	300
			7,400				7,400
20X8				20X8			
Jan 1	In arrear b/d	(5)	300	Jan 1	In advance b/d	(4)	100

Classwork 2

Received during the year 550	
Not yet receive at year beginning 350	at year end 200
Receive in advance at year beginning 600	at year end 300

---

[07] Donations (Income)

Any *donations* received are shown as income in the year that they are received. It can be from members or non-members.

The double entry of donations and entrance fees

Description	Double Entry
Donation received or entrance fee received	Dr: Cash Cr: Donation

[08] Receipts from specific activity

Any income and expenditure in a specific activity should be matched and shown in the Income & Expenditure a/c to reflect the profit of loss on the specific activity

Profit on activity will be treated as income	Actual Income > Actual Expense
Loss on activity will be treated as expenditure just like profit / loss on disposal.	Actual Income < Actual Expense

[09] Expenses Items

- Rent & Rates for organisation's premises
- Staff Salaries and Wages
- Maintenance Cost of Premises
- Other Running Expenses
- Loss on Fund Raising Activities

[10] Format of Final Accounts

Income & Expenditure a/c of a non-profit making organisation is used to reflect the actual income and expenditure in each financial year, whether the amounts of them are received or paid.

a) Format of Income & Expenditure a/c

Richard Club

Income & Expenditure a/c for the year ended 31 Dec 2007

Incomes	$	$
Subscription	100	
Life Membership Fee	200	
Bar Profit	100	
Entrance Fee	200	600
Expenditures		
Salaries	200	
Rent	100	300
Surplus		300

Incomes > Expenditure → Surplus

Expenditure > Incomes → Deficit

b) The format of the Balance Sheet of both will be identical. "Accumulated Fund (AF)" will replace the item "Capital" in the organisation.

| Opening AF | = | Opening Total Assets | - | Opening Total Liabilities |

[11] Bar Trading Account

A separate trading account is to be prepared for profit section such as bar, restaurant, refreshment, where is made for one kind of income.

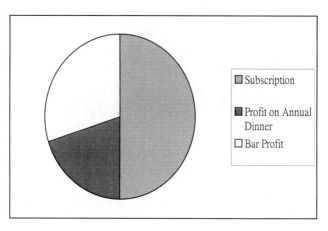

Incomes of the club and society

Classwork 3

Trial Balance as at 31 December 20X7

	Dr $.	Cr.$
Equipment	8,000	
Premises	29,000	
Subscriptions		6,000
Wages	4,000	
Furniture and fittings	5,000	
Rates and Rent	1,900	
Sundry expenses	600	
Accumulated fund 1 January 20X7		
Donations		300
Telephone ,Postage & Printing	500	
Ban k	2,000	
Bar purchases	9,500	
Creditors		1,000
Bar sales		14,000

Bar stocks 1 January 20X7	2,000

Classwork 4

The following information is also available: (i) Bar stocks at 31 December 20X7 $2,000. (ii) Provision for depreciation: sports equipment $1,700; Furniture and fittings $1,200.

Surplus =   >

Deficit =   <

[12] Different between Income & Expenditure a/c and Receipt & Payment a/c

	Receipts & Payment a/c	Income & Expenditure a/c
Nature	Record the cash in and out briefly	Calculate the surplus and deficit
Entry	Cash Book	Profit & Loss a/c

= 本章完　End of Chapter =

[01] Introduction

S K Ltd purchased and sold gold to earn profit.   The following is its trading during the month of Jan 20X8.

Gold bought (20X8)	Gold sold (20X8)
Jan 1 10 at $200 @	Jan 17 8 at $360@
Jan 12 10 at $240@	Jan 29 24 at $400@
Jan 28 20 at $250@	

What should be the closing stock at Jan 20X8?

Problem in making decision: Gold bought in different prices

Solution: Determine the ways to calculate the value of sock

Stock should not be valued at selling price as to do so would anticipate profit which has not yet been earned by the business.

所有 classwork 的工作紙儲存於 CD 內，可以預先列印。

[02] What is "stock"?

- for sale in the operating activities of the business; or
- In the process of production for sales such as work in progress; or
- In the form of materials for production such as raw materials.

[03] Net Realisable Values (NRV)

Saleable value	-	Expenses needed before completion of sales	=	NRV

[04] Measurement of Stock

Stock must be valued at the lower of its cost or net realizable value.

Classwork 1

Article	Different Categories	Cost
1	A	100
2	A	120
3	A	300
4	B	180

5	B	150
6	B	260
7	C	410
8	C	360
9	C	420

Category	Cost	Stock valuation
A	$(100+120+300) = $520	$100 x 3 = $300
B		
C		
Total		

The value of closing stock can be calculated under perpetual or periodic stock system.   However, they are still subject to the following adjustments
- Lower of cost or market value
- Movement of stock

[05]   Methods of stock valuation

a)   Cost of closing stock

$$\text{Cost of closing stock} = \text{Costs of purchase} + \text{Costs of Conversion} + \text{Other Costs}$$

<u>Costs of Purchase</u>
- Invoice price, Import duties, Trade discounts, other Taxes
- Transportation for materials, service and finished goods
- Other costs directly attributable to the buying

<u>Cost of Conversion</u> 加工成本
- Cost directly related to the units of production
- Production cost which incurred in converting materials into finished goods

Other Costs

They are included in the cost of stock in bringing stock to their present location and condition. For example, the costs of designing products for specific customers in the cost of stock in South Africa.

b) Examples of costs excluded from the cost of stock
- Abnormal (不可預期的) amounts of wasted materials, labour or other related production costs.
- Storage costs (treat as administrative expense in Profit & Loss a/c)
- Administrative Overheads (post in Profit & Loss a/c)
- Selling and transportation costs (treat as selling and distribution expense in Profit & Loss a/c)

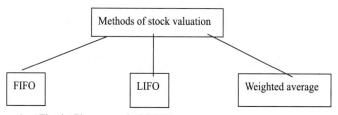

c) First in, First out method (FIFO)

FIFO is a method by which the first goods to be received are aid to be the first to be sold.

	Received	Issued	Stock after each transaction		
20X7 January				$	$
	10 at $300 each		10 at $300 each		3000
April	10 at $340 each		10 at $300 each	3000	6400
			10 at $340 each	3400	
May		8 x	2 at ____ each		
			10 at ____ each		4000
October	20 at $400 each		2 at ____ each		1,2000
			10 at ____ each		
			20 at ____ each		
November		2 x			
		10 x			
		12 x			
		24	8 at ____ each		

What is the Cost of Good Sold?

d) Last in, First out method (LIFO)

LIFO is a method by which the goods sold are said to have come from the last batch of goods to be received.

	Received	Issued	Stock after each transaction		
20X7				$	$
January	10 x $300		10 at $300 each		3000
April	10 at $340 each		10 at $300 each 10 at $340 each	3000 3400	6400
May		8 x	10 at ____ each 2 at ____ each		
October	20 at $400 each		10 at ____ each 2 at ____ each 20 at ____ each		
November		20 x 2 x 2 x <u>24</u>	8 at ____ each		

What is the Cost of Good Sold?

If we use the LIFO method, we assume that issues are always taken from the most recent deliveries.

e) Weighted average method (WAVO)

WAVO means the cost of units are averaged, and units used are charged at average cost.

	Received	Issued	Total Value	Number	Average
20X7			$		
January	10 at $30		10 x $300	10	

April	10 at $3400			20	
May		8 at $340		12	
October	20 at $400			320	
November		24 at $370		80	

Classwork 4

What is the Cost of Good Sold?

[06] Compare the different effects on the reported profit of the business

- With the WAVO, the items are issued at the average cost per unit, and the issue price is determined by dividing the total value by the number of units in stock
- This will tend to smooth out price fluctuations and the closing stock valuation will fall between that resulting from the FIFO and LIFO methods.
- In times of rising prices, the cost of good sold figure for the WAVO will higher than FIFO but lower than LIFO
- The profit will be lower than that calculated under FIFO but higher than that calculated under LIFO

Classwork 5

		FIFO		LIFO		WAVO
	$	$	$	$	$	$
Sales		50,000		50,000		50,000
Less: COGS						
Gross Profit						

[07] Loss of Stock

It may be required to calculate the loss of stock at a particular date sue to goods stolen by thieves or destroyed by fire.   Similar adjustments as made in the above example are needed to arrive at the amount of the loss.

=本章完 End of Chapter =

打劫都會因為分錢而出問題，試問合夥做生意時，如何可以避免此問題?

[01] Introduction

	Sole Trader Business	Partnership[1]
Owner / Owners	One	Two or more partners
Appropriation of net Profit	Not applicable as all the profit belongs to the sole	Shared by partners
Final Accounts	Trading a/c Profit & Loss a/c Balance Sheet	Trading a/c Profit & Loss a/c Appropriation a/c Balance Sheet

---

[1] Partnership Act 1890 defines a partnership as "the relationship which subsists between persons carrying on a business in common with a view to profit"

(a) The need for partnership

Partnerships are firms in which two or more people are working together to try to make profits.

(b) Nature of a partnership

Each partner, except for the limited partners, is liable to pay his share of partnership debts.

(c) Types of Partners

General Partners *(at lease 1 in partnership)*		*Sleeping / Silent Partners*

[02] Partnership Agreements

Written partnership agreements are preferable to mere oral agreements.

Underlying:

Capital Contribution	Profit / Loss sharing ratio	Interest on Capital

Interest on Drawings	Salaries

Sharing of Loss on realisation	Other special terms and conditions

(a) Profit or Loss sharing ratios

Profits or losses do not have to be shared in ratio to capital. It set in the partnership agreement such as Peter : Mary is 2:1.

(b) Interest on Capital

Interest on Capital can compensate for unequal investment of partners.

Capital a/c balance amount	x	Rate (%)	=	Interest on Capital

Classwork 1

Capital – A $20,000; B $15,000, Net Profit $12,000, Profit & Loss Sharing A to B : 1:1, Interest on Capital is 5%. What is the amount of total interest on capital?

所有 classwork 的工作紙儲存於 CD 內，可以預先列印。

(c) Interest on Current

Interest on Current can compensate for unequal current being contributed. It can be positive or negative.

Current a/c balance amount	X	Rate (%)	=	Interest on Current

Classwork 2

Example: Current a/c – A ($20,000); B $15,000; Interest on Current is 5%

(d) Salaries to partner

A salary can compensate a partner for extra duties or responsibilities.

Classwork 3

Lau & Cheung was the partners. Lau was entitled to a monthly salary of $2,400. The salary of Cheung was half that Lau from 1 Apr inwards, they receive new salary instead the old one. Lau monthly rate was increased by 10% upon the formation of the company while, Cheung's monthly rate was increased by 20% with effect from 1 July. What is their Partner's Salary?

(e) Interest on drawing

Interest on Drawing encourages partners to avoid excessive drawings.

Total Drawings	=	Cash Drawings	+	Stock Drawings

Total Drawings	X	Rate (%)	=	Interest on Drawings

Classwork 4

Example: Capital – A $20,000; B $15,000, Net Profit $50,000, Profit & Loss Sharing A to B : 1:1, Interest on Capital is 5%, Drawings – A:$2,000, B $4,000, Interest on Drawings is 1%, Partner Salary of both is $1000

Double entry	Dr	Cr
Interest on Capital Partner Salary Net Profit Sharing	Appropriation a/c	Partners' Current a/c

21

Drawings	Partners' Current a/c	Cash
Interest on Drawings	Partners' Current a/c	Appropriation a/c

[03] The final accounts

Profit and Loss a/c

- would be same as the sole traders'
- however, a Appropriation a/c will be opened to show the distribution of profits or losses among partners.

Richard Ltd

Appropriation a/c for the year ended 31 December 20X7

	$	$	$
Net Profit before Appropriation			10,000
Less: Appropriation			
Profit on Capital			
~ A	100		
~ B	200		
~ C	100	400	
Partner Salary			
~ A	100		
~ B	100		
~ C	100	300	
Interest on Current			
~ A	(200)		
~ B	200		
~ C	100	300	
Interest on Drawings			
~ A	200		
~ B	300		
~ C	100	(600)	400
Net Profit after Appropriation			9,600
Profit Sharing:			
~ A			3,200
~ B			3,200
~ C			3,200
			9,600

From Profit & Loss a/c

Balance Sheet
- also the same as the sole trader's in the part of Assets, Liabilities except Capital

Richard Ltd

Balance Sheet (extract) as at 31 December 20X7

Financed by:	A	B	C	
Capital a/c	10,000	10,000	10,000	30,000
Current a/c	8,000	(1,000)	5,000	12,000
	18,000	9,000	15,000	42,000

Long Term Liabilities

Loan				20,000
				62,000

[04] Capital Accounts and Current Account

The Capital account remain fixed the amount introduced as capitals.   The other items including interest on capital, salaries, share of profits and interest on drawing are entered in the partners' current accounts via the profit and loss appropriation account.

Capital a/c

	A $	B $	C $		A $	B $	C $
				Bal b/d			
Bal c/d				Cash			

Current a/c

	A $	B $	C $		A $	B $	C $
Bal b/d				Bal b/d			
Interest on Current Drawings				Interest on Capital			
Interest on Drawings				Interest on Current			
Net Loss				Partner Salary			
Bal c/d				Net Profit			
				Bal c/d			

=　本章完 End of Chapter =

[01] **Introduction**

Goodwill represents the **reputation** and **intangible** value of an **established** going concern.

- An intangible assets

  所有 classwork 的工作紙儲存於 CD 內，可以預先列印。

[02] **Goodwill exists because of:**

1. a good reputation – brand name
2. experienced, efficient & reliable employees – academic level
3. superior management team
4. good location
5. large, number of regular, customers
6. Going Concern – is it has profit growth?
7. Experience on the industry
8. The Possession of Favorable Contracts

[03] **Circumstances calling for the ascertainment of the value of goodwill are:**

1. Admission or withdrawal of Partners
2. Changes in Profit and Loss Sharing Ratio
3. Retirement or death of partner(s)
4. Amalgamation (combination) or transfer of business

Classwork 1

What is goodwill? Under what situation that will be record in a partnership? How to post in Balance Sheet?

[04] **Nature of Goodwill**

Total Price	-	Value of Assets	=	Goodwill

Classwork 2

You want to sell the business and assets and liabilities are listed:			
Buildings	225,000	Long Term Liabilities	100,000
Machinery	75,000	Creditors	30,000
Debtors	60,000	Bank Overdraft	45,000
Stock	40,000		

Mr. So paid $300,000 for the whole business, what is the amount of goodwill?

[05]   Types of Goodwill

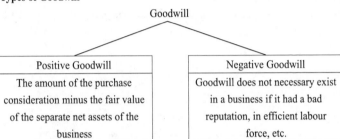

Positive Goodwill	Negative Goodwill
The amount of the purchase consideration minus the fair value of the separate net assets of the business	Goodwill does not necessary exist in a business if it had a bad reputation, in efficient labour force, etc.

[06]   Valuation

Goodwill is concerned with future profits. However, there is no single way of calculating goodwill to which everyone can agree but agreement reached between buyer and seller.

(i)   Average Sales Method

Average Yearly Sale for the past few years	x	Agreed Figure

- Usually for retail businesses

Classwork 3

It is agreed that the goodwill should be 3 times the average yearly sales for the past three years.   Calculate the value of goodwill:

Sales for the past three years:

20X5	$3,650,000
20X6	$5,000,000
20X7	$6,350,000

(ii)   Annual Fees Method

- For professional firms
- Based on gross annual income from fees, before charging expenses

Classwork 4

A CPA firm is selling its business.   It is asking a figure for goodwill, which is 2.5 times the average annual fees, received for the last two years. Calculate the goodwill:

20X6	$1,800,000
20X7	$2,200,000

(iii) Average Net Annual Profits Method

Average Net Profit for a no. of years	x	a stated amount

Classwork 5

Suppose goodwill is taken to be 4 times the average net profits for the past years. Calculate the goodwill.

20X5	$6,200,000
20X6	$6,900,000
20X7	$7,900,000

[07] Book- keeping Entries

(i) Goodwill Account Opened – stated in the Balance Sheet as "Intangible Fixed Assets"

Descriptions	Book-keeping Entries
Shared by EXISTING partners (the OLD profit sharing ratio)	**Dr.** Goodwill **Cr**. Partners' Capital Accounts

(ii) Goodwill Account **NOT** Opened

Descriptions	Book-keeping Entries
Shared by EXISTING partners (the OLD profit sharing ratio)	**Dr.** Goodwill **Cr**. Partners' Capital Accounts

Shared by partners (the NEW profit sharing ratio)	**Dr.** Partners' Capital Accounts **Cr**. Goodwill

Goodwill account is closed.

[08] Goodwill on Admission of partners

Classwork 6

R is admitted by bring in $4,000 cash.

Goodwill is agreed to be $10,000

Old sharing ratio: L(2), M(5), S(3)

New sharing Ratio: L(1), M(5), S(3), R(1)

(a)   Not to open a goodwill account – New partner pays his share of goodwill in cash into the business.   Old partners share the $ by increasing their capital, i.e., goodwill adjustment.

(b)   Not to open a goodwill account – New partner pays cash privately to old partners for his share of goodwill

[09]   Goodwill on withdrawal of partner

Classwork 7

A,B,C are partners share profits equally.
C withdraws from the partnership
Capital opening balances; $4,000 each
Goodwill is agreed to be $3,000 when C withdraws

(a)   When there is no Goodwill Account in the partnership:

Dr	Cr
Goodwill a/c	Capital a/c (old ratios)
Capital – partners gained	Capital – partners lost

Please prepare the accounts?
When C leaves the partnership, $_____ will be paid to him.

(b)   If there is a Goodwill Account in the partnership:
(i) Existing goodwill amount is the same ->
●   If a goodwill a/c exits with correct valuation of goodwill entered it, no further action
(ii) Goodwill is undervalued: e.g. goodwill already exist: $900
Please prepare the accounts?

(c)   Goodwill is overvalued:
e.g. Goodwill already exist: $3,600
Please prepare the accounts?

[10]   Goodwill on withdrawal and admission

Peter, John and May are partners sharing P & L in the ratio 1:2:3 respectively. May retired and June was admitted as a partner.   The new agreed P & L sharing ration of Peter, John and June are 2:3:5 respectively.   Goodwill was valued at $1,200.00.

Goodwill Account Opened			
Dr.	Goodwill	$1,200	
	Cr. Capital Accounts		
	- Peter (1/6)		$200
	- John (2/6)		$400
	- May (3/6)		$600

Goodwill Account NOT Opened			
Dr.	Goodwill	$1,200	
	Cr. Capital Accounts		
	- Peter (1/6)		$200
	- John (2/6)		$400
	- May (3/6)		$600
Dr.	Capital Accounts		
	- Peter (2/10)	$240	
	- John (3/10)	$360	
	- June (5/10)	$600	
	Cr. Goodwill		$1,200

Statement of Goodwill Adjustments Between Partners

		Peter	John	May	June	Total
		$	$	$	$	$
Goodwill was shared	1:2:3:-	200	400	600	-	1,200
Goodwill was written off	2:3:-:5	(240)	(360)	-	(600)	1,200
Net Adjustments		(40)	40	600	(600)	-

= 本章完 End of Chapter =

同學可於 CD 內開啓本課的 POWERPOINT，內有詳細的中文解釋如何處理商譽。

[01] Introduction

The assets and liabilities of a business may be revalued:

(1)   In admission or retirement of a partner, the assets and liabilities may be revalued to reflect the fair value of the business.

(2)   The partners change profit and loss sharing ratio

(3)   The business has be taken over by a limited company

(4)   Two or more partnerships are amalgamated

[02] Need for Revaluation

- The new partner would benefit from increase in value of assets before he joins the firm without paying anything
- The new partners would loss from decrease in value of assets with no revaluation takes place
- The assets of the partnership business should be revalued at the date of change so that any revaluation gain or loss of revaluation may be distributed to the partners before the change

[03] Book-keeping Entries

Description	Book-keeping Entries		
Transfer the reduction in the value of assets to Revaluation Account	**Dr**	Revaluation Account	
		Cr	Assets Account
Transfer the increase in the value of assets to Revaluation Account	**Dr**	Assets Account	
		Cr	Revaluation Account
Share Profit on Revaluation (Profit Sharing Ratio)	**Dr**	Revaluation Account	
		Cr	Partners' Capital Accounts
Share Loss on Revaluation (Loss Sharing Ratio)	**Dr**	Partners' Capital Accounts	
		Cr	Revaluation Account

Revaluation Account

REDUCTION In the value of assets	INCREASE In the value of assets
PROFIT ON REVALUATION	LOSS ON REVALUATION

Classwork 1

Land $5,000 → $10,000

所有 classwork 的工作紙儲存於 CD 內，可以預先列印。

Classwork 2

Car $15,000 → $11,000

[04] **Profit or Loss on Revaluation**

IF: New total valuation of assets > Old total valuation of assets = Profit on Revaluation

IF: New total valuation of assets < Old total valuation of assets = Loss on Revaluation

Classwork 3

Items	Dr	Cr
Showing gain on revaluation		
Showing loss on revaluation		
Showing an increase in total valuation of assets		
Showing a fall in total valuation of assets		

Classwork 4

W and Y share profit and losses in ratio 2/3, 1/3 at 31 December 20X7.   From 1 January 20X8, the ratio would be shared equally.

Balance Sheet as at 31 December 20X7			
Premises	65,000	Capital W	70,000
Fixtures	15,000	Capital Y	50,000
Stocks	20,000		
Debtor	12,000		
Bank	8,000		
The assets were revalued on 1 January 2007 to be:			
Premises $90,000, Fixtures $11,000, Provision for bad debts $3,000			

= 本章完 End of Chapter =

Chapter 8
Partnership - Dissolutions

[01] Introduction

Legally any change in the constitution of a partnership causes the old partnership to dissolve.

[02] Reasons

- The changes in the partnership due to the retirement or death of the partner(s).
- The bankruptcy of the business of the partnership
- The expiration of the partnership agreement.
- The business of the partnership was taken over by a limited company.

[03] The Procedures of Dissolution

Items	Process
Assets	• Sold to third party • Taken over by partners in their agreed price
Liabilities	• Settle it to someone who are not partners • Sold to third party
Loan from partners	Settle it to each partner as distinguished form capital.
Current a/c	Transfer the balances to their Capital Accounts.

[04] Book-keeping Entries

Steps	Descriptions	Book-keeping Entries Dr	Cr
(1)	Transfer provision accounts to relevant assets accounts e.g. Provision for Depreciation	Provision a/c	Assets a/c
(2)	Transfer the net book value of assets to Realisation Account, except bank a/c [All assets a/c will be closed]	Realisation a/c	Assets a/c
(3)	Assets sold to outsiders	Bank a/c	Realisation a/c
(4)	Liabilities which will be taken over by the partners	Liabilities a/c	Capital a/c
(5)	Net Assets taken over by partners in agreed price	Capital a/c	Realisation a/c
(6)	Payment to the Liabilities	Liabilities a/c	Bank a/c
(7)	Discount Received	Creditors	Realisation a/c

(8)	Cost of dissolution	Realisation a/c	Bank
(9)	Repayment to Loan	Loan a/c	Bank
(10)	Share the profit or loss on Realisation [close the Realisation account]	---	---
(11)	Settle the credit balances in partners' Capital Accounts to bank account	---	---

Classwork 1

Peter and Paul Co had land $100,000, car $12,000 with $3,000 accumulated depreciation, debtors $10,000, creditor $20,000 and bank $5,000. The partnership would be dissolute.

所有 classwork 的工作紙儲存於 CD 內，可以預先列印。

Classwork 2

The land sold at $200,000 and other assets were taken over by partners equal.

Classwork 3

Receive $9,000 from debtors for settlement the total debts.

Classwork 4

Paid $19,000 to creditor for settlement the total debts.

Classwork 5

Close all accounts.

(1) Balancing off the "Realisation a/c" and transfer the amounts to "Capital a/c"

(2) Balancing off the "Bank a/c" and transfer the amounts to related "Capital a/c"

(3) All the accounts should be closed and had not any balance carried down.

[05] Format of Realisation Account

Realisation a/c

	$	$		$	$
Book Value of FA		1,000	Capital a/c (take over)		
			~ A		1,000
			~ B		2,000
Book Value of CA		2,000	Assets sold		3,000
Bank		3,000	Creditors: discount		9,000
~ Cost of dissolution			received		

32

Profit on Realisation				
~A	3,000			
~B	3,000			
~C	3,000	9,000		
		15,000		15,000

The final balances on the partner's capital accounts should always equal the amount in the bank a/c from which they are to be paid.

[07] Conclusion

When a partnership is dissolved, all the assets are sold (or taken over by the partners individually), the liabilities are paid off and whatever remains is paid to the partners, the problems in dissolution accounting are:

(a) calculating the profit or loss on the realisation of the assets;

(b) arriving at the amounts due to or from the partners

Students must be thoroughly familiar with all aspects of the preparation of realisation account, bank account and partners' capital accounts.   All these accounts are key accounts of this topics and it's easy to prepare it by studying.

= 本章完 End of Paper =

這課題考試中出現機會很高，要掌握此課，唯一的方法是背熟第 31 頁的步驟。

[01] Introduction

Types of Business 企業類別	Owners 擁有人
Sole Proprietorship	Sole Trader 東主
Partnership	Partners 合夥人
Non-trading Organisation	Members 會員
Limited Companies	Shareholders 股東

At law, a limited company is **an artificial person** and is a separate **legal entity**.

- can own assets
- can bear liabilities in its own name
- can sue its own shareholders or vice versa.

"Limited Liability" means that

- Liability of a shareholder is limited to the nominal value of the shares he holds, whether fully paid or partly paid;
- The liabilities of the company itself are **still unlimited**

[02] Different between limited companies and partnerships

	Limited Companies	Partnerships
No. of owners	Public: 2 to unlimited Private: 1 to 50	2 to 20 (except in certain professional business such as CPA firms)
Amount of capital	Restricted to the amount of authorized capital	Stated in the partnership agreement
Distribution of Profit	Dividend	Share to each partners in their agreed ratio
Liabilities	Limited liability as the contribution amount on their shares held	Unlimited liability of each partner (except limited partners)
Management	Directors	Partners take part in the operations
Regulating Acts	Companies Acts	Partnership Act 1890
Rulers	Memorandum & Articles of Association	Partnership Agreements

Audit	Required	Optional
Profit distribution	Annual General Meeting	As per agreement

[03] Private and Public Limited Companies

	Private Limited Co.	Public Limited Co.
No of owners	1 to 50	At least 2
Share transfer	Agree with shareholders	Traded on HK Stock Exchange
Raising fund	ONLY from shareholders	From public
No of directors	1 to 50	At least 2
Scale	Small	Fever in number and generally much larger in size
Governed	Memorandum of Association, Articles of Association	

[04] Types of Shares

Ordinary Shares	Security or voting rights Have voting rights in general meetings Income 1. undistributed profits after senior claims have been made. 2. Any undistributed profits and all reserves belong to them. Amount of Capital 1. all surplus fund after prior claims 2. They own all reserves
Preference Shares	Security or voting rights Right to vote at general meeting only when it is proposed to change the legal rights of the shares Income 1. To paid before an Ordinary Share dividend may be paid. 2. Interest is not profit tax deductible. Amount of Capital A fixed amount per share per year

Rank of Liquidation

Liabilities (debentures)	⟹	Preferences Shares	⟹	Ordinary Shares

[05] **To increase the number of share**

(a) Issue to public

Dr	Cr
Bank	Application & Allotment
Application & Allotment	Ordinary Share Capital
	Share Premium

Classwork 1

On 1 Apr 20X8, Wai Wai Ltd issued 1,000,000 shares to the public at $0.7 (with $0.2 premium) each, payable in full upon application.

Please prepare the ledgers accounts.

所有 classwork 的工作紙儲存於 CD 內，可以預先列印。

(b) Issue to existing shareholders - Right Issues
- Give rights certificates to existing shareholders to buy shares at a specified price, and the specified price is normally the market price, so as to reward existing shareholders and to keep existing share-holdings unchanged (only when all shareholders used the right)
- Options of a shareholder:
  (a) ignore it
  (b) accept and buy shares
  (c) sell the right to other people for cash 投機

Debit (dr)	Credit (cr)
Bank	Ordinary Share Capital
Share Premium	

Classwork 2

On 20 Apr 20X8, the company received cash of $345,000 from an issue 300,000 shares through a right issue at a premium of $0.15 each.

(c) Bonus Issues
- Free shares to existing shareholders, no cash entries
- Why issue? Don't want to distribute cash as dividend
- Proportionate to the existing share-holding, e.g. 1 bonus share for every 5 shares held
- Market value of shares will be reduced from share premium

Debit (dr)	Credit (cr)
Share Premium	Ordinary Share Capital

Classwork 3

On 26 Apr 20X8, the company made a bonus issue of 1 share for every 5 ordinary shares in issue from its sharing premium.

[06] Reserves in a company

Reserves	=	Net Assets	-	Share Capital

	Sources	Distribution Methods
Revenue Reserves	Net Profit	Dividend
Capital Reserves	Issue share ate premium	Bonus Issue Discount on share issuing

[07] Provision

A charge against the profits, to provide for anticipated loss but the amount cannot be accurately determined.

(a) Provision for diminution in value of assets
- Provision for bad debts
- Provision for depreciation (Level 1)
- Provision for discount allowed

(b) Provision for liabilities or charge – loss or liability that is likely to be incurred, or is certain to be incurred but uncertain as to amount or date which it will arise
- Provision for long service payment
- Provision for warranty costs
- Provision for compensation

[08] Issue Price of raising capital

(a) Issue at Par

Issue Price	=	Par Value

Classwork 4

| Given Information: Number of shares proposed to issue = 10,000 |
| Issue Price = $100   Par Value = $100   No. of Application = 12,000 |

(b) Issue at a Premium

| Issue Price | - | Par Value | = | Share Premium |

Classwork 5

| Given Information: Number of shares proposed to issue = 10,000 |
| Issue Price = $120   Par Value = $100   No. of Application = 12,000 |

(c) Issue at a Discount

| Par Value | - | Issue price | = | Share Discount |

Classwork 6

| Given Information: Number of shares proposed to issue = 10,000 |
| Issue Price = $90   Par Value = $100   No. of Application = 12,000 |

Classwork 7

| Issue 1,000 ordinary shares of $0.5 each at a premium of $0.3 on 1 January 20X8. Total application received: 1,200 shares.   Unsuccessful application monies are refunded on 1 Feb 20X8. |

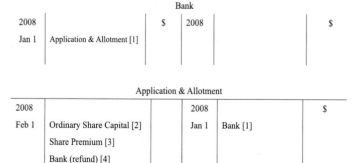

Bank

2008		$	2008		$
Jan 1	Application & Allotment [1]				

Application & Allotment

2008			2008		$
Feb 1	Ordinary Share Capital [2]		Jan 1	Bank [1]	
	Share Premium [3]				
	Bank (refund) [4]				

38

[09]   Case on Application

When a public company asks people to apply for its shares, it does not know how many shares will be applied for.   The company has to say how many shares it wants to issue.

over-subscription	Sometimes there will be more shares applied for than the company can issue.
under-subscription.	At other times, there will be fewer shares applied for than those issued

The number of application is named as "A". The number of issued is named as "I"

Relationship between "A" & "I"	Actual number Issued
A = I	I
A > I	I
A < I	A

Shares Issued at a Premium / Payable in full on application / over-subscription

	Dr	Cr
Received application money	Bank	Applications & Allotments
Shares allotted	Applications & Allotments	Ordinary Share Capital Share Premium
Refunds to Applicants	Applications & Allotments	Bank

Classwork 8

Issue 20,000 ordinary shares of $2 par at $4.   Application together with the necessary money, are received for 30,000 shares. Shares are allotted among the other applicants on a prorata basis, excess application monies are refunded

這課的重點是緊記發行股票的步驟，無論是 Journal 或"T"字帳，都要充份掌握。

= 本章完 End of Chapter =

[01]   Types of Debentures

Security or voting rights

1. No voting rights
2. A fixed charge on a specific asset or a floating charge on all assets
3. When liquidate, they entitle to the proceeds of their securities, if proceeds not sufficient, they become unsecured creditors.
4. If surplus, proceeds add to the assets for creditors.

Income

A fixed annual amount whether the company makes a profit or not.

Amount of Capital

A fixed amount per unit of debenture.

[02]   Issue of Debentures

a) In general case, the issue of debenture would issue at par.   That means the issuing price is same at the par value.

Issue at a par	Dr	Cr
Received application monies	Bank	Debenture Applicants
Debentures allotted	Debenture Applicants	Debenture

Classwork 1

On 1 January 20X8, 2,000 $100 10% debentures were issued at par, received in full on application.

所有 classwork 的工作紙儲存於 CD 內,可以預先列印。

b) If the company has the goodwill, the issue of debenture would issue at premium.   That means the issuing price is higher than the par value.

Issue at a premium	Dr	Cr
Received application monies	Bank	Debenture Applicants
Debentures allotted	Debenture Applicants	Debenture
		Share Premium

On 1 January 20X8, 2,000 $100 10% debentures were issued at 110, received in full on application.

c) In some special case such after SARS, the company want easy to issue the debenture, it will issue it at discount. That means the issuing price is lower than the par value.

The difference is so called "Debenture Discount" and it is one kind of financial expense.

Issue at a discount	Dr	Cr
Received application monies	Bank	Debenture Applicants
Debentures allotted	Debenture Applicants Debenture Discounts	Debenture
Amortisation of Discounts	Profit & Loss a/c	Debenture Discount

Classwork 3

On 30 Sep 20X8, 3,000 $100 10% debentures were issued at 90, received in full on application. Debenture Discount would be amortized among 5 years on a monthly basis.

[03] Issue of Debentures (over subscription)

If the company has the good reputation, the investors may have interest on its debentures. The quantity demanded of debentures may over than the quantity supplied of it.

a)

Issue at a par	Dr	Cr
Received application monies	Bank	Debenture Applicants
Debentures allotted	Debenture Applicants	Debenture
Refund	Debenture Applicants	Bank

Classwork 4

On 1 January 20X8, 2,000 $100 10% debentures were issued at 100, received 3,000 on application.   Applications were received debentures on 5 February.

b)

Issue at a premium	Dr	Cr
Received application monies	Bank	Debenture Applicants
Debentures allotted	Debenture Applicants	Debenture Share Premium
Refund	Debenture Applicants	Bank

Classwork 5

On 1 January 20X8, 2,000 $100 10% debentures were issued at 110, received in full on application.   Applications were received debentures on 5 February.

c)

Issue at a discount	Dr	Cr
Received application monies	Bank	Debenture Applicants
Debentures allotted	Debenture Applicants Debenture Discounts	Debenture
Amortisation of Discounts	Profit & Loss a/c	Debenture Discount
Refund	Debenture Applicants	Bank

Classwork 6

On 30 Sep 20X8, 2,000 $100 12% debentures were issued at 90, received 3,000 applications. Debenture Discount would be amortized among 5 years on a monthly basis.

= 本章完 End of Chapter =

Chapter 11
Limited Company – Final Accounts

[01] Introduction 簡介

For internal purposes, the final accounts of the limited companies are similar as those of the Sole Proprietorship and Partnership with the exception of one expense (e.g. Director's Fee) and the appropriation of net profit (e.g. dividend, transfer to / (from) reserves etc.)

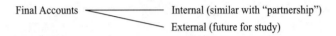

Final Accounts ──── Internal (similar with "partnership")
External (future for study)

[02] Special Types of Expenses 特別支出類別

(a) Debentures Interest / Loan Interest 債券利息 / 貸款利息

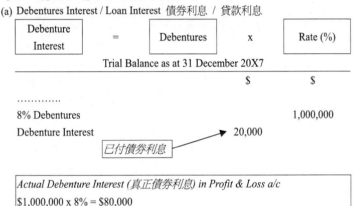

| Debenture Interest | = | Debentures | x | Rate (%) |

Trial Balance as at 31 December 20X7

	$	$
…………		
8% Debentures		1,000,000
Debenture Interest	20,000	

已付債券利息

Actual Debenture Interest (真正債券利息) in Profit & Loss a/c
$1,000,000 x 8% = $80,000

Accrued Debenture Interest (應付債券利息) in Balance Sheet (under the heading of "Current Liabilities")
Actual Debenture Interest – Paid Debenture Interest
$80,000 - $20,000 = $60,000

Classwork 1

Please prepare the Debenture Interest Account.

所有 classwork 的工作紙儲存於 CD 內，可以預先列印。

(b) Director's Fee 董事費 / Director's Emolument 董事酬金

Director's Fee and Director's Emolument are the salaries of the directors. It will be treated as one of administration expenses in the Profit & Loss a/c.

Classwork 2

Gross Profit = $100,000, Total Expenses = $78,000

The managing director was entitled to a salary equal to 10% of the net profit (after deduction of his salary). What is the salary of managing director?

(c) Preliminary Expenses / Formation Expenses 開辦費 / 創業費

Preliminary Expenses are the formation expenses of a limited company, including legal fees, professional charges etc.

[03] Appropriation of Net Profit 盈利的分配

(a) Dividends 股息

Net Profit will be distributed to its shareholders according to the level of net profit (盈利水平) and the dividend policy (派息政策) of the company.

Interim Dividend 中期股息

- paid dividend to the shareholders in the middle of the financial year.
- The amount will be subjected to the performance in the first half of the financial year.

Dr	Cr
Interim Dividend	Bank
Appropriation a/c	Interim Dividend

Classwork 3

Oct 1	Paid the interim dividend $5,000.
Dec 31	The account clerk prepares the final account.

Proposed Dividend (BS)/ Final Dividend (A) 擬派股息 / 末期股息

- will be subjected to the shareholders' approvals in the Annual General Meeting (AGM). 週年會員大會

Dr	Cr
Appropriation a/c	Proposed Dividend (Balance Sheet)

Classwork 4

- 5% 10,000 preference shares $1 each & Interim dividend is $300
- What should you to at the end of financial year at 31 Dec 20X7?

Transfer to / from Reserves 撥入或撥出儲備

It meet the future requirements such as Debenture Redemption.

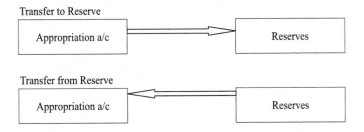

Transfer to Reserve

| Appropriation a/c | → | Reserves |

Transfer from Reserve

| Appropriation a/c | ← | Reserves |

Revenue Reserves can be transferred back to Appropriation a/c for Dividend purposes in the future financial period. (Same case of decrease of provision for doubtful debts)

Classwork 5

(1) Transfer $10,000 to General Reserve at 31 Dec 20X7.
(2) At the same date, Development Reserve has been decreased by $7,800.

[04] Appropriation Account

Richard Ltd

Appropriation a/c for the year ended 31 December 20X7.

	$	$	$
Net Profit before Appropriation			10,000
Less: Appropriation			
Interim Dividends			
~ Ordinary Dividend	100		

~ Preference Dividend	200	300	
Final Dividends			
~ Ordinary Dividend	200		
~ Preference Dividend	400	600	
Reserves			
Transfer to Redemption Reserve	1,000		
Transfer from General Reserve	(500)	600	1,500
Net Profit for the year			8,500
Retained Profit b/d			5,000
Retained Profit c/d			13,500

[05] Classification of Share Capital

(a) Authorised Share Capital (Nominal Capital)

The maximum amount of share that a company is authorised to issue as stated in the Memorandum of Association.

(b) Issued Share Capital

The amount of shares actually issued, it should not exceed authorised share capital

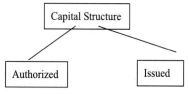

Classwork 6

Wai Wai Ltd had an authorised capital of 7,000,000 ordinary share of $0.5 each. On 1 April 20X8, it has already issued 3,000,000 ordinary shares and decided to issue all the remaining un-issued shares at $1 per share.

[06] Rights of shareholders

- To vote in the meeting to appoint directors
- To attend the Annual General Meeting
- To receive dividend (the amount given to shareholders as their share of profit)

[07] Other means of funding

Debentures	Reserves
• Long-term Loans	• Issue share of premium
• Fixed rate of interest payable	
• Fixed date of redemption	

[08] Balance Sheet

The Balance Sheet of the limited companies are similar 近似 as those of the Sole Trader 獨資經營 and Partnership 合夥經營 with the exception 除了 of the "Share Capital"股本 and "Reserves" 儲備 sections

Richard Happy Ltd

Balance Sheet (Extract) as at 31 December 2007

	$	$
Financed by:		
Authorised Share Capital 註冊股本		
2,000,000 Ordinary Shares of $2 each		XXX
1,000,000 Preference Share of $0.5 each		XXX
		XXX
Issued Share Capital 發行股本		
1,000,000 Ordinary Shares of $2 each, fully paid		XXX
500,000 Preference Share of $0.5 each		XXX
		XXX
Reserves 儲備		
Share Premium 溢價	XXX	
General Reserve 一般儲備	XXX	
Retained Profit (c/d) 留存利潤	XXX	
Proposed Dividend 擬派股息	XXX	XXX
		XXX

Classwork 7

(a) Net Profit before Appropriation $100,000

(b) Transfers to Debenture Redemption Reserve $50,000

(c) Transfer from General Reserve $20,000

(d) Interim Dividend (i) Ordinary $5,000 (ii) Preference $6,000

(e) Final Dividend (i) Ordinary $10,000 (ii) Preference $12,000

(f) Retained Profit b/d $50,000

= 本章完 End of Chapter =

[01]   **Manufacturing Account 製造成本帳**

Some of the firms manufacture the goods by themselves instead of purchases the goods from the suppliers.   In general, it can minimize the **cost of good sold** in the Trading a/c in the long run.   Therefore, **Manufacturing a/c** is prepared before the Trading a/c.

Comparison:

Account	Functions
Trial Balance	For checking contain balances of the Asset & Liability a/c
Trading a/c	To find out Gross Profit or Gross Loss
Profit & Loss a/c	To find out Net Profit or Net Loss
Appropriation a/c	To find out Retained Profit c/d
Manufacturing a/c	To find out Production cost

Production Cost of Goods Completed 完成品之生產成本

Manufacturing a/c

	$		$
Costs of goods manufactured	X	Value of goods manufactured	A

Trading a/c

	$		$
Value of goods manufactured	A		

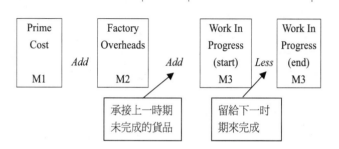

Classwork 1

	What is the production cost of goods completed?	$
M1	Prime Cost 主要成本	100
	Factory Overheads 間接費用	200
	Work in Progress – 1.1.20X7 半製成品	300
	Work in Progress – 31.12.20X7	200
	Rent 租金	300
	Sales 銷貨	1,000
T	Return Inwards	200

所有 classwork 的工作紙儲存於 CD 內，可以預先列印。

[02] Prime Cost 主要成本 (M1)

Prime Cost is the total direct cost in the progress of manufacturing that can be directly traced to the items manufactured. [Prime Cost 主要成本 → 3D]

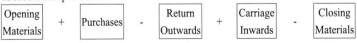

Direct Materials		Direct Labour		Direct Expenses
Raw Materials Consumed	*Add*	Workers who control the machine	*Add*	Royalties & Copyrights

(A) Direct materials are the costs of raw materials used in the production of goods such as rubber for sport shoes

| Opening Materials | + | Purchases | - | Return Outwards | + | Carriage Inwards | - | Closing Materials |

(B) Direct labour is the cost of manpower involved in the production line.   Wages paid to people not directly engaged in production but assisting in production are called " indirect labour costs"

(C) Direct expenses are the costs of expenses that can allocated to each goods such as trademark, patent, etc.

Which is the Prime Cost?

		$
M1	Opening Materials 期初原料	100
	Purchase – finished goods 購買完成品	200
	Purchase – raw materials 購買原料	300
	Return Outwards – finished goods 完成品退貨	100
	Return Outwards – raw materials 原料退貨	50
	Carriage Inwards – raw materials 原料購買運費	100
T	Closing Stock – finished 期末存貨	50
	Closing Stock – raw materials 期末原料	100
	Direct Manufacturing Wages 直接工資	400
	Factory Manager Wages 工廠經理工資	500
	Office Manager Salary 辦公室經理薪金	1,000
	Bonus of CEO 行政總裁獎金	10,000
	Royalties 專利權	600
PL	Advertising 廣告	700

所有屬於原料都是 M1

[03] Factor Overheads 間接製造成本 (M2)

Factory Overheads are all total indirect expenses in the process of manufacturing that cannot be traced to the items manufactured.

- Indirect materials – materials for repairs of machinery
- Indirect labour – wages to manager, cleaners
- Indirect expenses – provision for depreciation - machinery

Wages paid to people not directly engaged in production but assisting in production are called "indirect labour costs" such as wages paid to work supervisors storemen and time keepers

In the public examination and actual practice, some of the expenses items will be apportioned between the factory and office of the company. The former will go to the Manufacturing a/c while the latter will go to the Profit & Loss a/c.

Classwork 3

Rent & Rates 租金及差餉	$400,000
Insurance 保險	$15,000
Heating & Lighting 電費	$36,000
Telephone 電話費	$6,000

Provision for Depreciation 折舊準備

~ *office equipment* 辦公室設備	$1,000
~ *machinery* 機器	$2,000
~ *motor vehicle* 小型汽車	$4,000

Additional Information: 附加資料

- Insurance included $3,000 for the last years.
- Accrual telephone Expenses $500 本年度應付電話費開支$500
- 東主所住的租金 $10,000 Sole Trader's private rent $10,000
- Half time of motor vehicles was used in office.
- Expenses should be apportioned as follows:

	M2 Factory 工廠	PL Office 辦公室	Total $
Rent & Rates	3/4	1/4	
Heating & Lighting	3/4	1/4	
Insurance	5/6	1/6	
Telephone	1/2	1/2	
Provision for Depr.	---	---	

[04] Work in Progress 半製成品 (WIP) (M3)

Work in Progress is the semi-finished goods in the manufacturing process that cannot be sold. 半製成品是生產過程中未完成的產品。

The opening WIP is to be continuously worked on absorption further costs until it becomes completed goods. In fact, the opening WIP is "one of the resources used in the production of completed goods in the current period."

M1	+	M2	+	Opening WIP	-	Closing WIP	=	Production Cost

[05] Format of Manufacturing a/c 製造成本帳的格式

A firm engaged in manufacturing as well as trading activities will have a Trial Balance containing the balances of all accounts pertinent to the Manufacturing a/c and the Trading & Profit & Loss a/c. Needless to say, the trial balance will also contain balances of the Asset and Liability accounts.

M1 – Prime Cost / M2 – Overheads / M3 – WIP

Richard Ltd
Manufacturing a/c for the year ended 31 December 20X7

	$	$	$
Direct Materials 直接物資			
Opening Stock 期初存貨		1,000	
Purchase 購貨	500		
Return Outwards 購貨退出	(100)		
Carriage Inwards 購貨運費	200	600	
Closing Stock 期末存貨		(400)	1,200
Direct Labour 直接人工			
Manufacturing Wages 製造工資			2,000
Direct Expenses 直接費用			
Copyrights 版權			3,000
Prime Cost 主要成本			6,200
Factory Overheads 間接費用			
Indirect Wages , Rent & Rates, Electricity		3,000	
Provision for Depreciation – Machinery 折舊準備		2,000	5,000
			11,200
Work In Progress 1.1.2007 半製成品			500
			11,700
Work In Progress 31.12.2007 半製成品			(1,000)
Production Cost of Goods completed 生產成本 (轉移往購銷帳)			10,700

[06] **Cost of Goods completed transfer to Trading a/c**

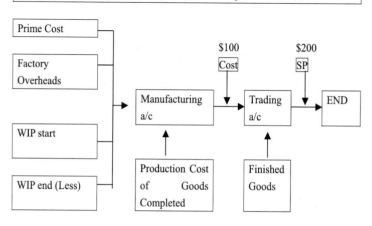

	$	$
Finished Goods Transfer to Trading a/c		2,500
生產成本移至購銷帳		
Sales 銷貨		20,000
Less: Cost of Good Sold 減：銷貨成本		
Opening Stock 期初存貨	1,000	
Production Cost 生產成本	2,500	
Purchase 購貨 ($20/unit)	4,000	
Closing Stock 期末存貨	(2,000)	5,500
Gross Profit 毛利		14,500
Add: Other Income 加：其他收入		
Interest 利息		1,000
		15,500
Less: Administration Expenses		
電費 Electricity	100	
Provision for depreciation - Vehicle	200	
Insurance 保險費	300	600
Selling & Distribution Expenses		
Sales Commission		
Net Profit 純利		14,900

[10] Treatment of closing stock 期末存貨處理

In a manufacturing business, there are three types of closing stock as follows:

Types	Final Accounts	Treatment of Closing Stock in Balance Sheet	
Raw Materials	Manufacturing a/c	Transfer at Production Cost of Goods Completed	
WIP Finished Good	Manufacturing a/c Trading a/c	Current Assets Stock	$
		~ raw materials	X
		~ WIP	X
		~ finished goods	X
			X

53

[11] Expenses charged to profit and loss a/c

a) Administration expenses

These expenses do not be affected by the sales volume.

- The salaries of directors, management and office staff;
- Rent and Rates;
- Insurance;
- Telephone and postage;
- Printing and stationery;
- Heating and lighting.

b) Selling and distribution expenses

There are expenses associated with the process of selling and delivering goods to customers.

- The salaries of a sales director and sales management.
- The salaries and commissions of salesmen.
- The traveling and entertainment expenses of salesmen.
- Marketing costs(e.g. advertising and sales promotion expenses).
- The costs of running and maintaining delivery vans.
- Discounts allowed to customers for early payment of their debt.
- Bad debts written off.
- Provision for Depreciation – Office Equipment

c) Finance expenses

For example: Interest on a loan; Bank overdraft interest.

= 本章完 End of Chapter =

[01] Introduction

Suspense account is an account showing a balance equal to the difference in the trial balance.

[02] Suspense account and the balance sheet

If the errors are not found before the final accounts are prepared, the suspense account balance will be included in the balance sheet.   When the balance is a credit balance, it should be included on liabilities side of the balance sheet.

[03] Correction of errors – one error only

Example 1: Assume that the error of $40 as shown in was found in the following year on 31 March 20X8.   The error was that the sales account was added up incorrectly and shown as $40 too little.

Suspense a/c

20X8		$	20X8		$
Mar 31	Sales	40	Jan 1	Balance b/d (bal)	40
		40			40

Sales

20X8		$	20X8		$
Mar 31			Jan 1	Suspense	40

Example 2: In another business, the trial balance on 31 Mar 20X8 had a difference of $168.   It was a shortage on the debit side of the trial balance.   A suspense account was opened.

Suspense a/c

20X8		$	20X8		$
Mar 31	Bal b/d (bal fig)	168	Jan 1	Trial Balance	168
		168			168

Example 3: On 31 Mar 20X8 the error was found. We had made a payment of $168 to K. Lee to close his account. It was correctly entered in the cash book, but it was not posted to K. Lee's account.

Suspense a/c

20X8		$	20X8		$
Mar 31	Bal b/d (bal fig)	168	Mar 31	K. Lee	168
		168			168

K. Lee

20X8		$	20X8		$
Mar 31	Suspense (correction)	168	Mar 31	Bal b/d	168
		168			168

[04] Correction of error – more than one error

Example 4: The trial balance as at 31 Mar 20X7 showed a difference of $77, being a shortage on the debit side. A suspense account was opened. On 28 February 20X8 all the errors from the previous year were found.

(i)   A cheque of $150 paid to Richard's had been correctly entered in the cash book, but had not been entered in Richard's account.

(ii)   The purchases account had been undercast by $20.

(iii)   A cheque of $93 received from K. San had been correctly in the cash book, but had not been entered in San's account.

Richard

20X8		$	20X8		$
Feb 28	Suspense (correction)	150	Feb	Bal b/d	150
		150			150

Suspense a/c

20X8		$	20X8		$
Feb 28	Bal b/d	77	Feb 28	Richard	150
	K San	93		Purchases	20
		170			170

Purchases

20X8		$	20X8		$
Feb 28	Suspense	20	Feb 28	Trading	20
		20			20

K. San

20X8		$	20X8		$
Feb 28	Bal b/d	93	Feb 28	Suspense	93
		93			93

[06] Profit adjustment

Example 5:

Net Profit $10,600

1. An amount paid for advertising $128 had been correctly entered in the Cash Book, but entered as $281 in the Advertising Account.
2. The balance of $347, on a customer's account, had been written off the Debtors Account, but the double entry had not been completed.
3. Discount allowed $223 had been correctly entered in Cash Book and in customer's account, but had been credited to the Discount Received Account.

Calculate the revised profit for the year ended 31 March 2006

	Dr	Cr	Increase	Decrease
(1)				
(2)				
(3)				

= End of Paper 本章完 =

[01] Introduction

Control Account is an account, which checks the arithmetical accuracy of a ledger.

Transactions

Credit Sales	Credit Purchase	Return Inwards	Return Outwards	Cash Purchase, Cash Sales, Paid Creditors, Received from Debtors, Cash drawings, Cash payment, Cash receipts	Bad Debt, Depreciation, Buy FA on credit, Corrections of errors ...	Sundry daily exp.
Sales Journal	Purchases Journal	Return Inwards Journal	Return Outwards Journal	Cash Book	The Journal	Petty Cash Book

"T" account in Sales Ledger/Purchase Ledger/General Ledger

Control accounts are sometimes known as "total accounts". It is an easy way to find out the balance of creditors and debtors. Thus, a control account for a Sale Ledger could be known either as a "sale ledger control account" or as a "total debtor account".

Debtors' Account   ⟶   Sales Ledger Control Account (SLC)

Transactions are summed up and recorded in respective control accounts

Creditors' Account	⟶	Purchase Ledger Control Account (PLC)

[02] Functions and advantages of control accounts

- To ensure the arithmetical accuracy of the entries
- For internal checking (department heads)
- To find out total debtors' and total creditors' balance easily without reference to sales ledger and purchase ledger accounts.
- To locate the difference with the aid of the control accounts when two sides of a trial balance do not agree
- Control accounts are kept in the general ledger
- Subsidiary ledger consisting of individual debtors/creditors accounts (individual ledger cards) is kept to support the control account.
- Control accounts can provide debtors/creditors balances quickly to facilitate the preparation of trial balance and balance sheet
- It shows a separation of duties in keeping the general ledger and subsidiary ledgers

[03] Information for control accounts

Sales Ledger Control	Source
1. Opening Debtors	The sum of balance b/d on all debtors.
2. Credit Sales	Total from Sale Journal / Sales Day Book.
3. Return Inwards	Total of Return Inwards Journal.
4. Cheque received	Bank Column of Cash Book on received side.
5. Cash received	Cash Column of Cash Book on received side.
6. Discount Allowed	Total Discount Allowed Column of Cash Book on received side.
7. Closing Debtors	The sum of balance c/d on all debtors

Sales Ledger Control a/c

	$		$
Bal b/d		Return Inwards	
Sales		Bank	
		Cash	
		Bad Debt	
		Discount Allowed	
		Bal c/d	

Purchase Ledger Control	Source
1. Opening Creditors	The sum of balance b/d on all creditors.
2. Credit Purchase	Total from Purchase Journal / Purchase Day Book.
3. Return Outwards	Total of Return Outwards Journal.
4. Cheque Paid	Bank Column of Cash Book on payments side.
5. Cash Paid	Cash Column of Cash Book on payment side.
6. Discount Received	Total Discount Received Column of Cash Book on payment side
7. Closing Creditors	The sum of balance c/d on all creditors.

Purchase Ledger Control a/c

	$		$
Bank		Bal b/d	
Cash		Purchase	
Discount Received			
Bal c/d			

Classwork 1

Debit Balance on 1 Jan 20X8	20,000
Total Credit Sales for the month	100,000
Cheques received from customers in the month	70,000
Cash received from customers in the month	12,000
Returns inwards from customers during the month	2,000
Debit balance on 31 Jan as extracted from the sales ledger	33,000
Total Cash Sales	50,000
Dishonor cheque	1,000
Bad Debt Recover	3,000
Bad Debt	4,000

所有 classwork 的工作紙儲存於 CD 內，可以預先列印。

> If the total of a control account are not equal and the entries made to it were correct, this shows that there is an error somewhere in the ledger.

Classwork 2

Credit Balance on 1 Jan 20X8	10,000
Total Credit Purchase for the month	100,000

Cheques paid to suppliers in the month	70,000
Returns outwards to suppliers during the month	2,000
Credit balance on 31 Jan as extracted from the purchase ledger	???
Cash Purchase	2,000
Discount Received	1,000

[04] Set-Off

A contra account whereby the same firm is both supplier and a customer and inter-indebtedness is set-off, will also need entering in the control accounts.

In General Ledger
Dr: Purchase Ledger Control   Cr: Sales Ledger Control

Classwork 3

Boz owed $10 to Apple.   Apple owed $7 to Boz. Accounts settled by transfer from one ledger to other $7. In the Boz's book

Classwork 4

What is the entry of Apple's book?

[05] Errors in sale ledger control

1. prime entry records are incorrect
2. adjustments not posted to sale ledger control account
3. the sales ledger control account may have been incorrectly balanced at the ended of the accounting period

[06] Errors in sale ledger

1. entries have been missed, or incorrectly made on individual customer accounts
2. individual debtor accounts might have been incorrectly balanced at the end of the accounting period
3. one or more debtor account balances might have been omitted from the listing of the sales ledger

Errors	Corrections	
	Dr	Cr
(a) Credit Sales Invoice omitted *(Increase the debtors ledger listing)*	SLC	Sales

(b)	Sales omitted in journal	SLC	Sales
	(Increase the debtors ledger listing)		
(c)	Sales of $12 incorrectly entered n sales journal as $10	SLC	Sales
	(Increase the debtors ledger listing)		
(d)	Sales journal undercast	SLC	Sales
	(The debtos ledger listing remain constant, because only the figure is wrong. This does not affect the debtor accounts, the figures for which are entered individually.)		
(e)	Sales of $12 incorrectly entered as $10 in sales accounts	Suspense	Sales
(f)	Sales understated in SLC	SLC	Suspense
(g)	Sales of $12 incorrectly debited to debtor account as $10.	No Entry	
(h)	Balance of sales has been incorrectly put to the wrong side of the trial balance	Suspense	---
(i)	Balance of debtors has been incorrectly put to the wrong side of the trial balance	No entry	
(j)	Balance of SLC a/c has been incorrectly put to the wrong side of the trial balance	---	Suspense

= 本章完 End of Chapter =

這課題是第二級最艱深的一課

Chapter 15
Accounting Ratios & the Interpretation of Accounts

[01] Introduction

The following bases may be used to compare the accounting ratios of current financial period.

1. The accounting ratios of last financial period
2. The budgeted accounting ratios of current financial period
3. The accounting ratios of other companies in the same industry

Can be used to ascertain the performance of a business in a particular period, or to provide a basis for comparing the results of the current year with those of previous or for comparing the results of different companies in the same industry.

[02] Use of Ratio Analysis

Ratio analysis is a process of evaluating the relationship between the component parts of financial statements to obtain a better understanding of a firm's performance and financial position.

By ratio analysis, the performance of the firm can be evaluated:

1.	profitability,	Level 2
2.	liquidity	Level 2
3.	management efficiency	Level 2 & Level 3
4.	investment return	Level 3
5.	financial stability	Level 3

Parties interested in analyzing financial statements

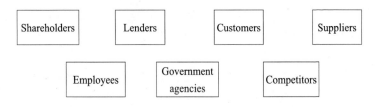

Classwork 1

Why they want to know the ratios?

所有 classwork 的工作紙儲存於 CD 內，可以預先列印。

Classwork 2

> If commission is 5% of profit after commission is deducted, and profit before commission is $16,800, then commission is ?

[02] Markup & Margin

$$\text{Mark up} = \frac{1}{3}$$

$$\text{Margin} = \frac{1}{3+1} \Rightarrow \frac{1}{4}$$

- With same numerator of markup 分子相同
- Denominator = mark up's denominator + numerator 分母會加上分子

Classwork 3

> If mark up is 25%, what is margin?

Classwork 4

> If margin = 20%, what is mark up?

[03] Calculating Missing Figures

In examination, the main use of Ratios is……… calculating missing figure.

Classwork 5

> All goods in a firm have the same rate of markup and ignore wastage and theft of goods.   What is the amount of sales? [Hints: Open the trading account]
>
> Figures for 20X7
>
> | | |
> |---|---|
> | Stock 1.1.20X7 | $400 |
> | Stock 31.12.20X7 | $600 |
> | Purchase | $5,200 |
> | A uniform mark-up rate 20% | |

[04] Calculation of Accounting Ratios – Profitability

To measure the performance of management and to identify whether the company is a worthwhile investment opportunity relative to its competitors

(i)   Gross Profit Ratio / Gross Profit Margin

$$\text{Margin} = \frac{\text{Gross Profit}}{\text{Net Sales}} \times 100\%$$

- It shows the amount of GP for every $100 of sales

- Sales increased -→ GP % fallen
- It indicates the efficiency of cost control & pricing policy of company
- A higher gross profit ratio may indicate the goods are being sold at higher
- Sometimes the company may lower its gross profit margin in order to stimulate sales (law of demand)
- A low gross profit ratio may not result in a small figure of gross profit unless a small volume of sales accompanies it.
- It can defect the errors and fraud affecting the items in the trading account
- The staff of Inland Revenue to defect the suppression of sales uses it.

Classwork 6

Why the gross profit ratio would be decrease?

(ii) Net Profit Ratio

$$\text{Net Profit Ratio} = \frac{\text{Net Profit}}{\text{Net Sales}} \times 100\%$$

- It indicates the efficiency of expenses control
- Sometimes a high gross profit ratio with a low net profit ratio may indicate too much money is spent on the operating expenses. Investigation should be made on the operating expenses to find out which expense occupied a high proportion.

Classwork 7

High margin, low Net Profit Ratio.   What does it mean?

(iii) Return on Capital Employed

$$\text{ROCE} = \frac{\text{EBIT}}{\text{Total Assets} - \text{CL}} \times 100\%$$

- The most important ratio of all
- It indicates the efficiency of management in generating profits from the capital employed
- The higher the ratio, the more profit the company gained
- It enables the profitability of the company comparable with the other investment opportunities.

Classwork 8	Company A	Company B
Net Profit for the year	50	65
Share Capital	100	150
General Reserve	60	50
Retained Profit	40	75
Debentures	50	50
Calculate the return on total capital employed? Which company is more profitable?		

- In exam, use the method stated by the examiner

[05] Calculation of Accounting Ratios – Liquidity

A company's liquidity is its ability to pay its debt by having enough cash or liquid assets easily available

(i) Current Ratio / Working Capital Ratio

$$\text{Current Ratio} = \frac{\text{CA}}{\text{CL}} : 1$$

- A high current ratio may indicate the company has sufficient current assets to meet its current liabilities
- A low current ratio may indicate the company may face the liquidity problems (not be able to meet its current liabilities)

(ii) Quick ratio (Acid test ratio)

$$\text{Quick Ratio} = \frac{\text{CA} - \text{stock}}{\text{CL}} : 1$$

- to indicate the ability of a firm to meet it's pay out of its Quick Assets[1].
- A high quick ratio may indicate the company has sufficient liquid assets to meet its current liabilities

Classwork 9

		ABC Ltd	DEF Ltd
Current Assets:	Stock	300	300
	Debtors	200	200
	Bank	100	100

---

[1] quick assets mean the assets which can be converted into cash very quickly

Current Liabilities: Creditors	400	200
Working Capital Ratio & Quick Ratio		

- If the quick ratio is too high, the company may be holding too much idle liquid assets such as debtors or cash
- A low quick ratio may indicate the company may face the liquidity problems
- The difference between current ratio and quick ratio may indicate the proportion of stock holding. The greater the difference may indicate the company is holding too much stock.

Classwork 10

What is the solution to improve the liquidity position?

[06] Calculation of Accounting Ratios –Activity & Efficiency

Efficiency ratios are used to assess the efficiency of financial management of a company such as stock control, credit control and asset management.

(i) Stock Turnover

$$\text{Stock Turnover Rate} = \frac{\text{Cost of Good Sold}}{\text{Average Stock}} = \text{X times}$$

$$\text{Stock Turnover Rate} = \frac{\text{Average Stock}}{\text{Cost of Good Sold}} \times 365 \text{ days} = \text{X days}$$

$$\text{Stock Turnover Rate} = \frac{\text{Average Stock}}{\text{Cost of Good Sold}} \times 52 \text{ weeks} = \text{X weeks}$$

$$\text{Stock Turnover Rate} = \frac{\text{Average Stock}}{\text{Cost of Good Sold}} \times 12 \text{ months} = \text{X months}$$

- It shows how well the firm is managing to do these things.
- Any increase in stock or slowdown in sales will show a lower ratio.
- If the opening stock is not given then the closing stock is assumed to be the average stock.

Classwork 11

What does it mean the low stock turnover?

How to improve the stock turnover?

1. Stock Turnover = 5 times	2. Stock Turnover = 12imes
Opening Stock = $600	Opening Stock = $2,000
Closing Stock = $900	Closing Stock =
Purchase =	Purchase = $16,000

A firm should try to do the following

(i) Keep its stock to as low a figure as possible without losing profitability.

(ii) Sell its goods as quickly as possible.

(ii) Turnover of Debtors / Credit Period Allowed to Debtors

$$\text{Turnover of Debtors} = \frac{\text{Trade Debtors}}{\text{Credit Sales}} \times 365 \text{ days} = \text{X days}$$

$$\text{Turnover of Debtors} = \frac{\text{Trade Debtors}}{\text{Credit Sales}} \times 52 \text{ weeks} = \text{X weeks}$$

$$\text{Turnover of Debtors} = \frac{\text{Trade Debtors}}{\text{Credit Sales}} \times 12 \text{ months} = \text{X months}$$

- This ratio assesses how long it takes for our debtors to pay us
- The shorter the debtors collection period the better
- Sometimes, the company may offer generous credit terms to its customers in order to stimulate sales
- The higher the ratio, the worse we are at getting our debtors to pay on time the lower the ratio, the better we are at managing our debtors.

Why the firm wants debtors pay faster?

(iii) Turnover of Creditors / Credit Period Received from Creditors

$$\text{Turnover of Creditors} = \frac{\text{Trade Creditors}}{\text{Credit Purchases}} \times 365 \text{ days} = \text{X days}$$

$$\text{Turnover of Creditors} = \frac{\text{Trade Creditors}}{\text{Credit Purchases}} \times 52 \text{ weeks} = X \text{ weeks}$$

$$\text{Turnover of Creditors} = \frac{\text{Trade Creditors}}{\text{Credit Purchases}} \times 12 \text{ months} = X \text{ months}$$

- This ratio shows how long it takes a firm on average to pay its suppliers
- The lower creditors turnover the better

Classwork 15

Why for a firm to take longer time to pay its suppliers?

(iv) Assets Turnover Ratio

$$\text{Assets Turnover Ratio} = \frac{\text{Net Sales}}{\text{Total Assets}} = X \text{ times}$$

- It indicates the efficiency of utilization of total assets in generating revenue
- A low capital (assets) turnover may indicate:
  1. an inefficient management and utilization of assets
  2. an excessive investment in assets

Classwork 16

How to improve the capital (assets) turnover?

(v) Debtors : Creditors Ratio

The ratio compares how much credit is allowed by the business to its customers (debtors), and how much credit has been received by the business from its suppliers (creditors).

[07] Limitations of Ratio Analysis

- Based on historical cost - it may not be relevant for predicting future performance.
- Different firms may adopt different accounting policies which make comparisons difficult such as depreciation on SL or RB
- The uniqueness of a particular year or a particular firm has to be considered before a meaningful comparison can be made.

- Changes in the value of money and the price levels may weaken the validity of comparisons of ratios computed for different years such as the appreciation of RMB since 2006 and inflation in 2008
- Since the financial statements are prepared at a date, they do not disclose the short-term fluctuations during the accounting period.

[08] Overtrading

- It occurs when there is insufficient control over working capital resulting in insufficient liquid funds to meet the demands of creditors
- It is generally the result of too fast sales growth in relation to level of debtors, creditors and stock
- If it occur: raise long-term finance, or cut-back on the expansion

[09] Sample of interpretation of Accounting Ratios

1.

Profitability	Firm A	Firm B	Remarks
Gross Profit Margin			
Net Profit Margin			
Return on Capital Employed			

Analysis:

1. *By comparing their ROCE, Firm A is (more profitable), (50)% compared with (29)% of Firm B. → Firm A is (more efficient) in utilizing the capital to generate revenue.*
2. *Firm A has (higher profit margin) of the sales, (42)% compared with (24)%. → use high (profit margin policies)*
3. *Firm A's NP% (drop more) firm GP% than Firm B's, from (42)% to (15)%, compared with (24)% to (10)% of Firm B. → Firm B is (more efficient) / has better control of.*

2.

Liquidity	Firm A	Firm B	Remarks
Current Ratio			
Quick Ratio			

Analysis:

1. Current Ratio: both companies are in acceptable range, (2) to 1 and (3.1) to 2. → able to repay (debts) when it is done and have sufficient (working capital) for further investment and operation. → Firm B has (more cash) pay back its debts as its ratio is higher (1.3) to 1 compared with (0.75) to 1 of Firm B.

2. Quick Ratio: Firm A has higher ability to pay (urgent payment), Firm B <1 → (poor liquidity position)

3. Compare 2 ratios: Firm B has higher current ratio but lower quick ratio → (build up) in stock level / (accumulation) of stock → (trading slows down)

3.

Efficiency	Firm A	Firm B	Remarks
Stock Turnover Ratio			
Debtors Collection Period			
Creditors Repayment Period			

Analysis:

1. *Firm A has (better stock level control), it rate is (3.4) times than (2.9) times of Firm B.*
2. *Firm B has higher risk of (stock accumulation) → refer to quick ratio*
3. *We should also consider:*
- *(the nature of business)*
- *(industry average)*
4. *Firm B has better_____, it takes (1.8) months to collect its debt while Firm A takes (2.2) months.*
5. *Firm A can settle its debt in _____. (5.1) months compared with _____ months → better _____*
6. *If the period is too long → risk of _____ and _____*
7. *Compare two ratios: Firm A can have more _____ in the company*

[09] Examination Techniques

1. When calculating the ratios, you should:
- State the formula
- Substitute the suitable figure into the formula
- Computer the result
- Add the correct unit

2. When interrupting the ratios, you should:
- State out which firm/year is better in profitability or liquidity
- Support your answer with ratios
- Compare the ratios and give simple explanation

= End of Chapter = 本章完

Chapter 16
Incomplete Records & Single Entry

[01] Introduction

The owner may only maintain the information as follows by single entry system:

Cash / bank receipts and payments	Fixed assets

Debtors	Creditors

Reasons for recording in single entry system
- Limited resources for some small business
- Firms would enter details of a transaction once only
- Many of the businesses fail to record every transaction resulting in incomplete records.

Disadvantages:
1. The records of business transactions are incomplete
2. The final accounts cannot be produced quickly and accurately because of the absence of nominal accounts
3. The final results are not reliable because the operating results of the company are estimated based on the increase or decrease of capital in the period.
4. Errors and frauds may occur and are difficult to detect

[02] Calculation of Net Profit / (Loss) from Statement of Affairs

a) Statement of Affairs is same as balance sheet

Statement of Affairs (incomplete records)	=	Balance Sheet (double entry system)

Richard Ltd
Statement of Affairs as at 31 December 2007

Assets	$	$
Fixed Assets	4,000	
Currents Assets	5,600	9,600
Less: Liabilities		
Current Liabilities	1,500	
Long Term Liabilities	3,600	5,100
Capital as at 31 Dec 2007		4,500

b) Find Net Profit by the movement of Capital a/c

Capital a/c

	$		$
Drawings	2,500	Bal b/d	1,000
		Bank	1,500
Bal c/d	2,000	Net Profit	

c) Find Net Loss by the movement of Capital a/c

Capital a/c

	$		$
Drawings	500	Bal b/d	1,000
Net Loss		Bank	1,500
Bal c/d	1,000		

Classwork 1

The cash balance as at 1 Jan 20X7 and 31 December 20X7 are $70 and $80 respectively. Total cash sales and cash purchase are $1,000 and $800 respectively. What is amount of cash drawings?

Classwork 2

Brenda Lau's bank account for the year 20X7 is as follows:

	$		$
Balance 1.1.20X7	1,890	Cash withdrawn from bank	5,400
Receipts form debtors	44,656	Trade creditors	31,695
Loan from T. Ho	2,000	Rent	2,750
		Rates	1,316
		Drawings	3,095
		Sundry expenses	1,642

Records of cash paid were: sundry expenses $122, trade creditors $642.
Cash Sales amounted to $698, cash drawings were $5,289.

The following information is also available:

	31.12.20X6($)	31.12.20X7($)
Cash in hand	48	?

(a) What is the amount of Cash in the closing balance?.

[03] Preparation for trading, profit and loss a/c from incomplete records

a) The steps in the preparation of final accounts from incomplete records

    (i)   Drawing up an opening balance sheet / statement of affairs
- To find out the amount of opening capital

    (ii) Preparing summary cash and bank accounts
- To find out the amount of payment to specific item or;
- To find out the amount of receipt from specific item or;:
- To find out the amount of cash / bank drawings or;
- To find out the amount of closing balance

    (iii) Scheduling creditors
- To find out the amount of credit purchases or;
- To find out the amount of payment to creditors or;
- To find out the amount of closing creditors

    (iv) Scheduling debtors
- To find out the amount of credit sales or;
- To find out the amount of receipts from debtors or;

- To find out the amount of closing debtors

(v) To find out the amount of resalable stock and (or) closing stock

(vi) Making the year end adjustments (provision, reserves, prepayments)

(vii) Preparing the final accounts

b) The structure of Trading a/c

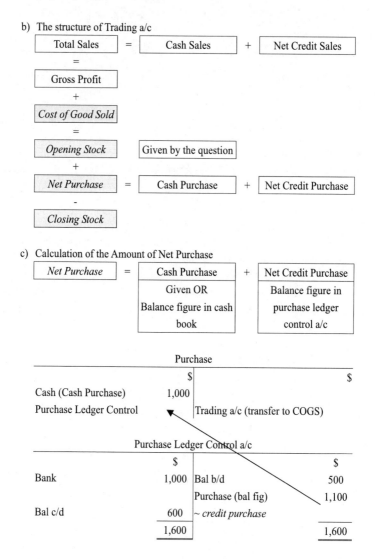

| Total Sales | = | Cash Sales | + | Net Credit Sales |

=

Gross Profit

+

Cost of Good Sold

=

Opening Stock   Given by the question

+

Net Purchase = Cash Purchase + Net Credit Purchase

-

Closing Stock

c) Calculation of the Amount of Net Purchase

Net Purchase	=	Cash Purchase	+	Net Credit Purchase
		Given OR		Balance figure in
		Balance figure in cash		purchase ledger
		book		control a/c

Purchase

	$		$
Cash (Cash Purchase)	1,000		
Purchase Ledger Control		Trading a/c (transfer to COGS)	

Purchase Ledger Control a/c

	$		$
Bank	1,000	Bal b/d	500
		Purchase (bal fig)	1,100
Bal c/d	600	~ credit purchase	
	1,600		1,600

d) Calculation of the Amount of Total Sales

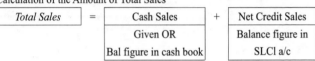

Total Sales	=	Cash Sales	+	Net Credit Sales
		Given OR		Balance figure in
		Bal figure in cash book		SLCl a/c

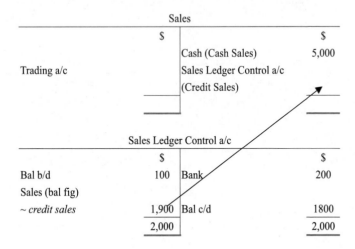

Sales

	$		$
		Cash (Cash Sales)	5,000
Trading a/c		Sales Ledger Control a/c	
		(Credit Sales)	

Sales Ledger Control a/c

	$		$
Bal b/d	100	Bank	200
Sales (bal fig)			
~ credit sales	1,900	Bal c/d	1800
	2,000		2,000

Classwork 3

The following figures have been extracted from the records of Brenda who does not keep a full record of his transactions on the double entry system:

1 April	20X7	Debtors	2,760
1 April	20X7	Creditors	1,080
1 April	20X7	Stock	2,010
31 March	20X8	Debtors	3,090
31 March	20X8	Creditors	1,320
31 March	20X8	Stock	2,160

All goods were sold on credit and all purchases were made on credit. During the year ended 31 March 20X8, cash received from debtors amounted to $14,610 whereas cash paid to creditors amounted to $9,390.

(i) Calculate the amount of sales ad purchases for the year ended 31 March 20X8.

(ii) Draw up the trading account for the year ended 31 March 20X8.

[04] Calculation of the amount of cash misappropriated

The amount of the cash of organisation may be misappropriated.

| Sales Ledger Control a/c in order to calculate the "cash and cheque received" | → | Cash Book The amount of cash misappropriated |

Sales Ledger Control a/c

	$		$
Bal b/d	200	Bad Debts	100
Sales	1,800	Cash (bal fig)	
		Bal c/d	400
	2,000		2,000

Cash Book

	$		$
Bal b/d	500	Cash Misappropriated (bal)	
Sales Ledger Control		Bal c/d	600

Classwork 4

1. Bank Deposit from Cash a/c is $972,000
2. The following amounts were paid out of cash: salary $15,000, sundry expense $24,000, drawings $20,000.
3. Cash Receipt from Debtor is $56,000
4. A refund of $12,000 cash from a supplier for goods returned has not been reordered.
5. Cash sales is $1,201,000

How much does cash misappropriated?

[05] Summary - How to find the missing figure?

Missing figures	Methods
Sales	● Total debtors a/c / Sales Leger Control a/c ● Trading a/c and margin
Purchases	● Total creditors a/c / Purchase Ledger Control a/c ● Trading a/c and markup
Stock Loss	● Stock Turnover Ratio ● Trading a/c
Capital	● Statement of Affairs ● Return on Capital Employed
Net Profit	● Statement of Affairs ● Return on Capital Employed ● Different between opening and closing capital
Cash misappropriated	● Cash Book

= 本章完 End of Chapter =

[01]　補充 – 計數機特別使用法 (加數計算機如 CASIO JS-140)

1.　<u>SPECIAL FUNCTIONS OF CALCULATIOR</u>

1.1　Example 1　　　A, B and C are in partnership sharing profits and losses in the ratio of 3: 2: 1 respectively.

Net Profit is $34,680

CA	3	4	6	8	0

Your calculator will display →　　K　　Which represents constant.

<u>A's share of Net Profit</u>

3　=　Your calculator will display →　　| 17,340 |

<u>B's share of Net Profit</u>

| 2 | = | Your calculator will display →　　| 11,560 |

<u>C's share of Net Profit</u>

| 1 | = | Your calculator will display →　　| 5,780 |

<u>Check Total</u>

| GT | Your calculator will display →　　| 34,680 |

1.2　<u>Example 2</u>

Capital – A : $20,000
Capital – B : $36,800
Capital – C : $19,700

<u>Interest on Capital is 12% p.a.</u>

CA	.	1	2	X	X

Your calculator will display →　　K　　which represents constant.

<u>A's Interest on Capital</u>

2	0	0	0	0	0	=

Your calculator will display →　　| 2,400 |

80

B's Interest on Capital

| 3 | 6 | 8 | 0 | 0 | = |

Your calculator will display → 4,416

C's Interest on Capital

| 1 | 9 | 7 | 0 | 0 | = |

Your calculator will display → 2,364

Total Interest on Capital

| GT | Your calculator will display → 9,180

[02] 出題新方向 – 混合式題目

| Provision for Depreciation | + | Capital & Revenue Expenditure |

| Accounting Ratio | + | Comments |

| Cash Book | + | Bank Reconciliation Statement |

| Partners Retirement | + | Partners Admission |

| Accounting Ratios | + | Incomplete Records |

| Correction of Errors | + | Control a/c |

| Stock Valuation | + | Stock Loss |

| Manufacturing a/c | + | Issue of Shares & Final Accounts |

| Accounting Ratios | + | Cost of Stock Destroyed |

[03] 各課題的問題範例

課題		問與答
原始紀錄	問	State the differences between trade discounts and cash discounts.
	答	1. *計算方法：Trade discounts directly deduct from the list price of the goods or services before trading.* 2. *入帳處理：It usually records in the sales day book or purchase day book* 3. *目的：It is offered when the large quantity of goods sold or maintaining the relationship with customers* 1. *計算方法：Cash discounts usually record in the (three column) cash book.* 2. *目的：It aims to make payment early* 3. *入帳方法：It will post to discount allowed a/c or discount received a/c.*
折舊準備	問	What are the causes of depreciation?
	答	*Physical deterioration, amortization, damage in unusual, obsolescence*
呆帳準備	問	What is the difference between bad debts and provision for doubtful debts?
	答	*Refer the text book chapter 2.*

82

統制帳	問	State briefly any two advantages of maintaining control accounts
	答	*Bookkeeping errors can be located quickly* *Total figures for creditors and debtors can be quickly produced* *Separation of duties reduces the possibly of fraud 欺詐.* *The control a/c provides an automatic check of the postings in the ledger.*
錯誤更正	問	List all the types of errors that not affecting trial balance agreement.
	答	*1) Errors of Commission 2) Errors of Principles 3) Errors of Omission 4) Errors of Original Entry 5) Compensating Errors 6) Complete reversal of entries*
現金簿	問	What is dishonoured cheque? State the causes.
	答	*It means that the cheque cannot be presented because:* ● *Insufficient fund* ● *Wrong in the drawer's signature* ● *Wrong in the date of the cheque*
存貨價值	問	State three methods for calculating the value of stock. What means by NRV?
	答	*FIFO, LIFO, WAVO* *NRV: Saleable value – expenses incurred before selling*
合夥經營	問	What elements should be included in a partnership agreement?
	答	*Capital Contribution, Profit/Sharing Ratio, Interest on Capital, Interest on Drawings, Partner Salaries, Arrangement for the admission, Arrangement for a partner retires*
	問	What happens if there is no partnership agreement?
	答	*Profit and Loss share equally, No interest on capital, No interest on drawings, No partners salaries, Any partner withdraw or join need to be agreed with all partners*
	問	In what situations that partnership goodwill should be revalued?

	答	*Change in Profit / Loss sharing ratio, admission of new partners, retirement of existing partners, sales of the partnerships*
會計比率與 分析	問	List three ratios that can reflect the profitability for a company.
	答	*Gross Profit Ratio, Net Profit Ratio, Return on Capital Employed*

[04]熱門課題分析

(a) Correction of Errors and Suspense a/c
這課題因出題範圍極度廣泛,故比較困難,在試卷的出題率謹次於決算帳,考生須特別留意。

答題步驟
(I) 首先詳細閱讀題目,小心完成 Journal 關於錯誤更正的部份
(II) 按已做的錯誤更正,編制 Suspense a/c [注意: 只有部份錯誤涉及 Suspense a/c]
(III) 按照已完成的 Journal,修正新的 Net Profit / Gross Profit

Example
Return Outwards of $200 to Richard had been recorded only in the personal account ("ONLY" is the key word)

處理步驟:
(I) 先列出正確的入帳方法,然後再列出題目的錯誤入帳法,最後提出改正方法

	Dr	Cr
正確	Creditor - Richard	Return Outwards
錯誤	Creditor - Richard	---
改正	*Suspense*	*Return Outwards*

(II) 根據步驟(I),編制 Suspense a/c

Suspense

	$		$
Return Outwards	200		

(III) Return Outwards 會影響 Gross Profit 及 Net Profit，Net Profit 調整如下：

Brenda Company

Statement of adjusted Net Profit as at …….

	$
Net Profit before adjustment	xxx
Add: Understate the Return Outwards	200
Net Profit after adjustment	xxx

(b) Accounting Ratios

此課題有兩種兩核方法：

1. 利用題目的資料直接計算比率，然後加以評論分析 (比較容易)
2. 推理式試題，利用題目中的 Ratios 及零碎資料完成不完整的帳戶，例如：Trading, Profit & Loss Appropriation a/c 及 Balance Sheet

直算式答題技巧：

應分別列出方程式，再列出計算過程，最後才寫出答應。

方程式			計算過程	2007 年	2008 年
ROCE	=	Net Profit			
		Capital Employed			
…					
….					
…					

推理式答題技巧

作答推理式題目，應細題目中的零碎資料及比率，然後逐一將遺失數字尋找出來，那當然，尋找的次序是不能有誤，而要準確地找出推理模式，相信要靠經驗的累積了。

Example:

Prepare the trading and profit and loss account and calculate the gross profit.

Margin = 20% Sales = $20,000 Return Outwards = $2,000

Stock Turnover rate = 4 times Opening stock = $30,000

Trading, Profit & Loss a/c for the year ended 31 December 20X7

			$		$		$
Sales							200,000
Less:	Cost of Good Sold						
	Opening Stock				30,000		
	Purchases	[5]	182,000				
	Return Outwards		(2,000)	[4]	180,000		
	Closing Stock			[3]	(50,000)	[2]	160,000
Gross Profit						[1]	40,000

STEP

[1] Gross Profit = Sales x margin = $200,000 x 20%

[2] COGS = Sales – Gross Profit = 200,000 – 40,000 = 160,000

[3]

$$\text{Stock Turnover Rate} = \frac{\text{Cost of Good Sold}}{\text{Average Stock}}$$

$$4 = \frac{160,000}{(30,000 + CS) \times 0.5}$$

$$4 = \frac{160,000}{15,000 + 0.5CS}$$

$$60,000 + 2CS = 160,000$$

$$2CS = 100,000$$

$$CS = 50,000$$

[4]

OS + Net Purchase – CS	=	Cost of Good Sold
30,000 + Net Purchase – 50,000	=	160,000
Net Purchase – 20,000	=	160,000
Net Purchase	=	180,000

[5]

Total Purchase – Return Outwards	=	Net Purchase
Total Purchase – 2,000	=	180,000
Total Purchase	=	182,000

(c) Provision for Depreciation
- 首要注意該資產所使用的折舊方法，直線法 / 餘額遞減法
- 注意該資產的購買日期及年結日期，可能不足一年，要清楚計算月份，建議以數手指方法最可靠
- 另外，一些必要開支，如運輸費 Transportation，應包括於固定資產帳內。計算 Depreciation 時必須包括該開支
- Disposal 的步驟及日期亦應特別注意

(d) Control a/c

要於 Control a/c 中取得較高分數，考生必須知道統制帳中各金額的來源，詳情可見後表。

Sales Ledger Control a/c		Purchase Ledger Control a/c	
Information	Sources	**Information**	Sources
Opening Balance	List of all debtor accounts drawn up at the start of the period	**Opening Balance**	List of all creditor accounts drawn up at the start of the period
Credit Sales	Sales day book	**Credit Purchase**	Purchase day book
Return Inwards	Return Inwards Day Book	**Return Outwards**	Return Outwards Day Book
Cheque received	Bank a/c	**Cheques paid**	Bank
Cash received	Cash a/c	**Cash paid**	Cash a/c
Discount Allowed	Discount column in Cash Book	**Discount Received**	Discount column in Cash Book
Bad Debts	Journal	---	---
Closing Balance	List of all debtor accounts drawn up at the end of the period	**Opening Balance**	List of all creditor accounts drawn up at the end of the period

[11]　後記

● 充份準備才能使你的壓力下降！
● Sosir2006@hotmail.com 全天候支援